Wojbor A. Woyczyński

A First Course in Statistics for Signal Analysis

Birkhäuser
Boston • Basel • Berlin

Wojbor A. Woyczynski
Department of Statistics and
Center for Stochastic and Chaotic Processes
in Science and Technology
Case Western Reserve University
10900 Euclid Avenue
Cleveland, OH 44106
U.S.A.

Cover Design by Joseph Sherman.

Mathematics Subject Classification : 60-01, 60G12, 60G15, 60G35, 62-01, 62M10, 62M15, 62M20

Library of Congress Cataloging-in-Publication Data
Worczynski, W. A. (Wojbor Andrzej), 1943-
A first course in statistics for signal analysis / by Wojbor A. Woyczynski.
p. cm.
Includes bibliographical references and index.
ISBN 0-8176-4398-2
1. Signal processing-Statistical methods. I. Title.
TK5102.9.W69 2005
621.382'2'05195-dc22 2005053570

ISBN-10: 0-8176-4398-2 e-ISBN: 0-8176-4516-0
ISBN-13: 978-0-8176-4398-0

Woyczynski: A First Course in Statistics for Signal Analysis

First Indian Reprint 2009

ISBN 978-81-8489-143-0

This edition is published by Springer (India) Private Limited,
A part of Springer Science+Business Media, Registered Office: 906-907, Akash Deep Building, Barakhamba Road, New Delhi - 110 001, India.

Printed in India by Rashtriya Printers, Delhi.

This book is dedicated to my children:
Martin Wojbor, Gregory Holbrook, and Lauren Pike.
They make it all worth it.

This book is dedicated to my children,
Martin Wolbor, Gregory Holbrook, and Lauren Pike.
They make it all worth it.

Contents

Introduction

This book was designed as a text for a first, one-semester course in statistical signal analysis for students in engineering and physical sciences. It had been developed over the last few years as lecture notes used by the author in classes mainly populated by electrical, systems, computer and biomedical engineering juniors/seniors and graduate students in sciences and engineering who have not been previously exposed to this material. It was also used for industrial audiences as educational and training material and for an introductory time series analysis class.

The only prerequisite for this course is a basic two- to three-semester calculus sequence; no probability or statistics background is assumed except the usual high school elementary introduction. The emphasis is on a crisp and concise but fairly rigorous presentation of fundamental concepts in the statistical theory of stationary random signals and relationships between them. The author's goal was to write a compact but readable book of approximately 200 pages countering the recent trend towards fatter and fatter textbooks.

Since Fourier series and transforms are of fundamental importance in random signal analysis and processing, this material is developed from scratch in Chapter 2 emphasizing the time domain vs. frequency domain duality. Our experience showed that although harmonic analysis is normally included in the calculus syllabi, students' practical understanding of its concepts is often hazy. Chapter 3 introduces basic concepts of probability theory, law of large numbers and the stability of fluctuations law, and statistical parametric inference procedures based on the latter.

In Chapter 4 the fundamental concept of a stationary random signal and its autocorrelation structure is introduced. This time domain analysis is then expanded to the frequency domain by discussion in Chapter 5 of power spectra of stationary signals. How stationary signals are affected by their transmission through linear systems is the subject of Chapter 6. This transmission analysis permits a preliminary study of the issues of designing filters with the optimal signal-to-noise ratio; this is done in Chapter 7. Chapter 8 concentrates on Gaussian signals where the autocorrelation structure completely determines all the statistical properties of the signal. The text concludes, in Chapter 9, with a description of algorithms for computer simulations of stationary random signals with given power spectrum density. The routines are based on the general spectral representation theorem for such signals, which is also derived in this chapter.

The book is essentially self-contained, assuming the indispensable calculus background mentioned above. Complementary bibliography, for readers who would like to pursue the study of random signals in greater depth, is described at the end of this volume.

Some advice to students using this book: The material is deliberately written in a compact, economical style. To achieve understanding needed for independent solving of the problems listed at the end of each chapter in the Problems and Exercises sections, it is not sufficient to read through the text in the manner you would read through a newspaper or a novel. It is necessary to look at every single statement with a "magnifying glass" and to decode it in your own technical language so that you can use it *operationally* and not just be able to talk about it. The only practical way to accomplish this goal is to go through each section with pencil and paper, explicitly completing, if necessary, routine analytic intermediate steps that were omitted in the exposition for the sake of the clarity of the presentation of the bigger picture. It is the latter that the author wants you to keep at the end of the day; there is no danger in forgetting all the little details if you know that you can recover them by yourself when you need them.

Finally, the author would like to thank Professors Mike Branicky and Ken Loparo of the Department of Electrical and Computer Engineering, and Professor Robert Edwards of the Department of Chemical Engineering of Case Western Reserve University for their kind interest and help in development of this course and comments on the original version of this book. My graduate students Alexey Usoltsev and Alexandra Piryatinska also contributed to the editing process and I appreciate the time they spent on this task. Partial support for this writing project from the Columbus Instruments International Corporation of Columbus, Ohio, Dr. Jan Czekajewski, President, is here also gratefully acknowledged.

Three anonymous referees spent considerable time and effort trying to improve the original manuscript. I appreciate their help. Special thanks go to Professor Craig Zirbel of the Department of Mathematics and Statistics at Bowling Green State University, who carefully read the manuscript with his usual attention to detail, aesthetics, and pedagogical worthiness of the exposition. His sage advice was incorporated almost without exception in the final version of the book.

Notation

Note: This is to be used only as a guide and not as a set of formal definitions.

BW_n	equivalent-noise bandwidth of the system
$\mathrm{BW}_{1/2}$	half-power bandwidth of the system
$\mathbf{C}$	the set of all complex numbers
$\mathrm{Cov}(X,Y) = \mathbf{E}[(X - \mathbf{E}X)(Y - \mathbf{E}Y)]$	covariance of X and Y
δ_{mn}	Kronecker delta, $= 0$ if $m \neq n$ and $= 1$ if $m = n$
$\delta(x)$	Dirac delta "function"
E_x	energy of signal $x(t)$
$\gamma_X(\tau) = \mathbf{E}(X(t) - \mu_X)(X(t+\tau) - \mu_X)$	autocovariance function of a stationary signal $X(t)$
$\mathbf{E}(X)$	expected value (mean) of random quantity X
$F_X(x)$	cumulative distribution function (c.d.f.) of random quantity X
$f_X(x)$	probability density function (p.d.f.) of random quantity X
$\varphi_{X,Y} = \mathbf{E}(XY)$	correlation of X and Y
$\phi_X(\tau) = \mathbf{E}(X(t)(X(t+\tau))$	autocorrelation function of a stationary signal $X(t)$
$h(t)$	impulse response function of a linear system
$H(f)$	transfer function of a linear system
$\lvert H(f)\rvert^2$	power transfer function of a linear system
$L_0^2(\mathbf{P})$	space of all zero-mean random quantities with finite variance
$m_\alpha(X) = \mathbf{E}\lvert X\rvert^\alpha$	αth absolute moment of random quantity X

$\mu_k(X) = \mathbf{E}(X^k)$	kth moment of random quantity X
$N(\mu, \sigma^2)$	Gaussian (normal) probability distribution with mean μ and variance σ^2
P	period of a periodic signal
$\mathbf{P}(A)$	probability of event A
PW_x	power of signal $x(t)$
$Q_X(\alpha) = F_X^{-1}(\alpha)$	α's quantile of random quantity X
R	resolution
$\mathbf{R}$	the set of all real numbers
$\rho_{X,Y} = \mathrm{Cov}(X,Y)/(\sigma_X\sigma_Y)$	correlation coefficient of X and Y
$\mathrm{Std}(X) = \sigma_X = \sqrt{\mathrm{Var}(X)}$	the standard deviation of random quantity X
$S_X(f)$	power spectral density of stationary signal $X(t)$
$S_X(f)$	cumulative power spectrum of stationary signal $X(t)$
$\boldsymbol{X} = (X_1, X_2, \ldots, X_d)$	a random vector in dimension d
T	sampling period
$x(t), y(t)$, etc.	deterministic signals
x_{av}	time average of signal $x(t)$
$\mathrm{Var}(X) = \mathbf{E}(X - \mathbf{E}X)^2 = \mathbf{E}X^2 - (\mathbf{E}X)^2$	the variance of random quantity X
$\lfloor a \rfloor$	"floor" function, the largest integer not exceeding number a
$u(t)$	Heaviside unit step function; $u(t) = 0$ for $t < 0$, and $= 1$ for $t \ge 0$
$W(n)$	discrete-time white noise
$\mathcal{W}(n)$	cumulative discrete-time white noise
$W(t)$	continuous-time white noise
$\mathcal{W}(t)$	the Wiener process
$x(t) * y(t)$	convolution of signals $x(t)$ and $y(t)$
$X(f), Y(f)$	Fourier transforms of signals $x(t)$ and $y(t)$, respectively
X, Y, Z	random quantities (random variables)
z^*	complex conjugate of complex number z; i.e., if $z = \alpha + j\beta$, then $z^* = \alpha - j\beta$
$\langle .,. \rangle$	inner (dot, scalar) product of vectors or signals
$\Leftrightarrow$	if, and only if
$:=$	is defined as

A First Course in Statistics for Signal Analysis

1

1

Description of Signals

Signals are everywhere, literally. The universe is bathed in the background radiation, the remnant of the original Big Bang, and as your eyes scan this page, a signal is being transmitted to your brain where different sets of neurons analyze it and process it. All human activities are based on processing and analysis of sensory signals, but the goal of this book is somewhat narrower. The signals we will be mainly interested in can be described as *data* resulting from quantitative measurements of some physical phenomena, and our emphasis will be on data that display *randomness* that may be due to different causes, such as errors of measurements, algorithmic complexity, or the chaotic behavior of the underlying physical system itself.

1.1 Types of random signals

For the purposes of this book, signals will be functions of real variable t interpreted as time. To describe and analyze signals we will adopt the functional notation: $x(t)$ will denote the value of a nonrandom signal at time t. The values themselves can be real or complex numbers, in which case we will symbolically write $x(t) \in \mathbf{R}$, or, respectively, $x(t) \in \mathbf{C}$. In certain situations it is necessary to consider vector-valued signals with $x(t) \in \mathbf{R}^d$, where d stands for the dimension of the vector $x(t)$ with d real components.

Signals can be classified into different categories depending on their features. For example, there are the following:

- *Analog signals* are functions of continuous time and their values form a continuum. *Digital signals* are functions of discrete time dictated by the computer's clock, and their values are also discrete and dictated by the resolution of the system. Of course, one can also encounter mixed-type signals which are sampled at discrete times but whose values are not restricted to any discrete set of numbers.

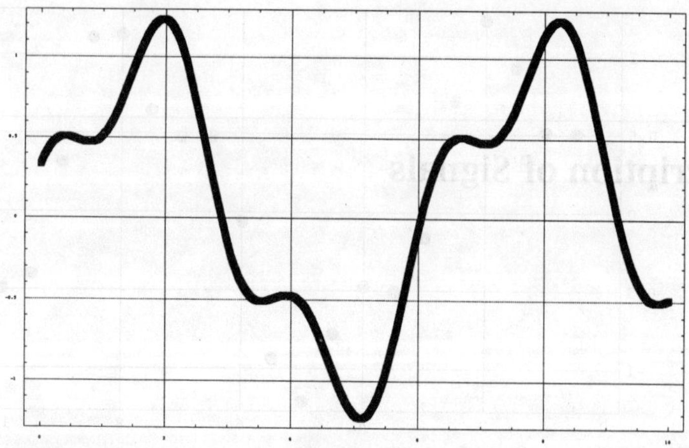

Fig. 1.1.1. Signal $x(t) = \sin(t) + \frac{1}{3}\cos(3t)$ [V] is analog and periodic with period $P = 2\pi$ [s]. It is also deterministic.

- *Periodic signals* are functions whose values are periodically repeated. In other words, for a certain number $P > 0$, we have $x(t + P) = x(t)$ for any t. The number P is called the *period of the signal*. *Aperiodic signals* are signals that are not periodic.
- *Deterministic signals* are signals not affected by random noise; there is no uncertainty about their values. *Stochastic* or *random signals* include an element of uncertainty; their analysis requires use of statistical tools, and providing such tools is the principal goal of this book.

For example, signal $x(t) = \sin(t) + \frac{1}{3}\cos(3t)$ [V] shown in Figure 1.1.1 is deterministic, analog, and periodic with period $P = 2\pi$ [s]. The same signal, digitally sampled during the first five seconds at time intervals equal to 0.5 s, with resolution 0.01 V, gives tabulated values:

t	0.5	1.0	1.5	2.0	2.5	3.0	3.5	4.0	4.5	5.0
$x(t)$	0.50	0.51	0.93	1.23	0.71	−0.16	0.51	−0.48	−0.78	−1.21

This sampling process is called the *analog-to-digital conversion*: given the *sampling period* T and the *resolution* R, the digitized signal $x_d(t)$ is of the form

$$x_d(t) = R\left\lfloor \frac{x(t)}{R} \right\rfloor \quad \text{for } t = T, 2T, \ldots, \tag{1.1.1}$$

where the (convenient to introduce here) "floor" function $\lfloor a \rfloor$ is defined as the largest integer not exceeding real number a. For example, $\lfloor 5.7 \rfloor = 5$, but $\lfloor 5.0 \rfloor = 5$ as well.

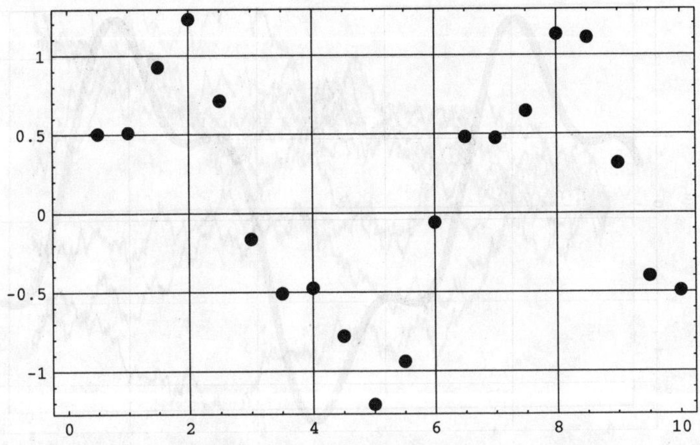

Fig. 1.1.2. Signal $x(t) = \sin(t) + \frac{1}{3}\cos(3t)$ [V] digitally sampled at time intervals equal to 0.5 s with resolution 0.01 V.

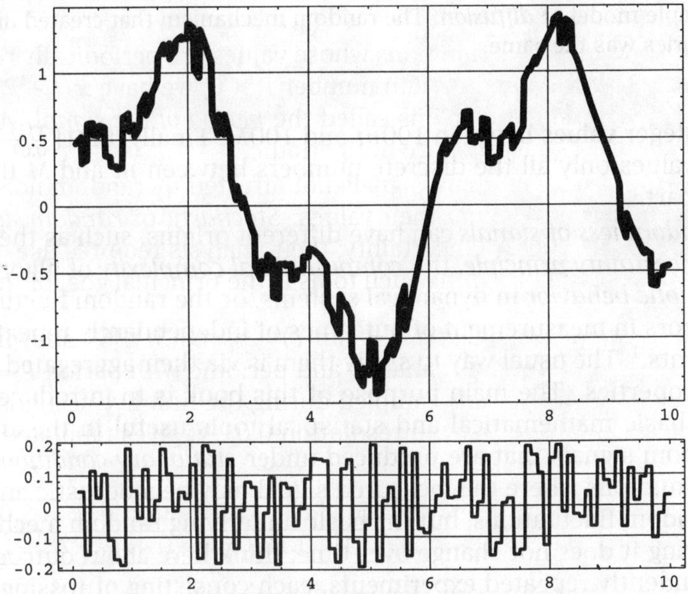

Fig. 1.1.3. Signal $x(t) = \sin(t) + \frac{1}{3}\cos(3t)$ [V] in the presence of additive random noise with average amplitude of 0.2 V. The magnified noise component itself is pictured underneath the graph of the signal.

Note the role the resolution R plays in the above formula. Take, for example, $R = 0.01$. If the signal $x(t)$ takes all the continuum of values between $m = \min_t x(t)$ and $M = \max_t x(t)$, then $\frac{x(t)}{0.01}$ takes all the continuum of values between $100m$ and $100M$, but $\lfloor \frac{x(t)}{0.01} \rfloor$ takes

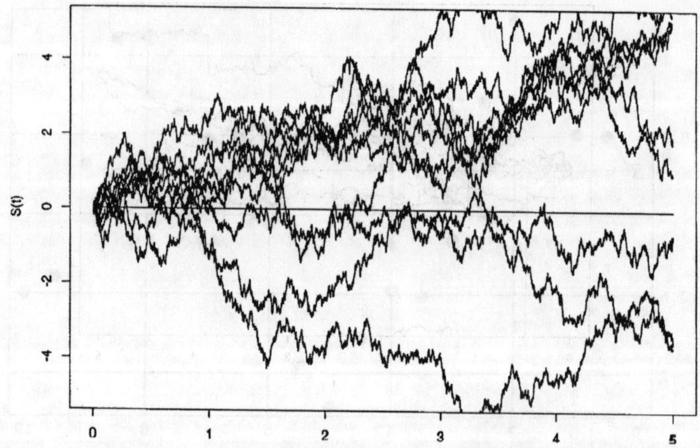

Fig. 1.1.4. Several computer-generated trajectories (sample paths) of a random signal called the *Brownian motion* stochastic process or the *Wiener stochastic process*. Its trajectories, although very rough, are continuous. It is often used as a simple model of *diffusion*. The random mechanism that created different trajectories was the same.

only integer values between $100m$ and $100M$. Finally, $0.01\lfloor \frac{x(t)}{0.01} \rfloor$ takes as its values only all the discrete numbers between m and M that are 0.01 apart.

Randomness of signals can have different origins, such as the quantum *uncertainty principle*, the *computational complexity* of algorithms, the *chaotic behavior* in dynamical systems, or the random fluctuations and errors in measurement of outcomes of independently repeated experiments.[1] The usual way to study them is via their aggregated statistical properties. The main purpose of this book is to introduce some of the basic mathematical and statistical tools useful in the analysis of random signals that are produced under *stationary conditions*, that is, in situations where the measured signal may be stochastic and contain random fluctuations, but the basic underlying random mechanism producing it does not change over time; think here about outcomes of independently repeated experiments, each consisting of tossing a single coin.

At this point, to help the reader visualize the great variety of random signals appearing in the physical sciences and engineering, it is worthwhile to review a gallery of pictures of random signals, both experimental and simulated, presented in Figures 1.1.4–1.1.8. The captions explain the context in each case.

[1] See, e.g., M. Denker and W. A. Woyczyński, *Introductory Statistics and Random Phenomena: Uncertainty, Complexity, and Chaotic Behavior in Engineering and Science*, Birkhäuser Boston, Cambridge, MA, 1998.

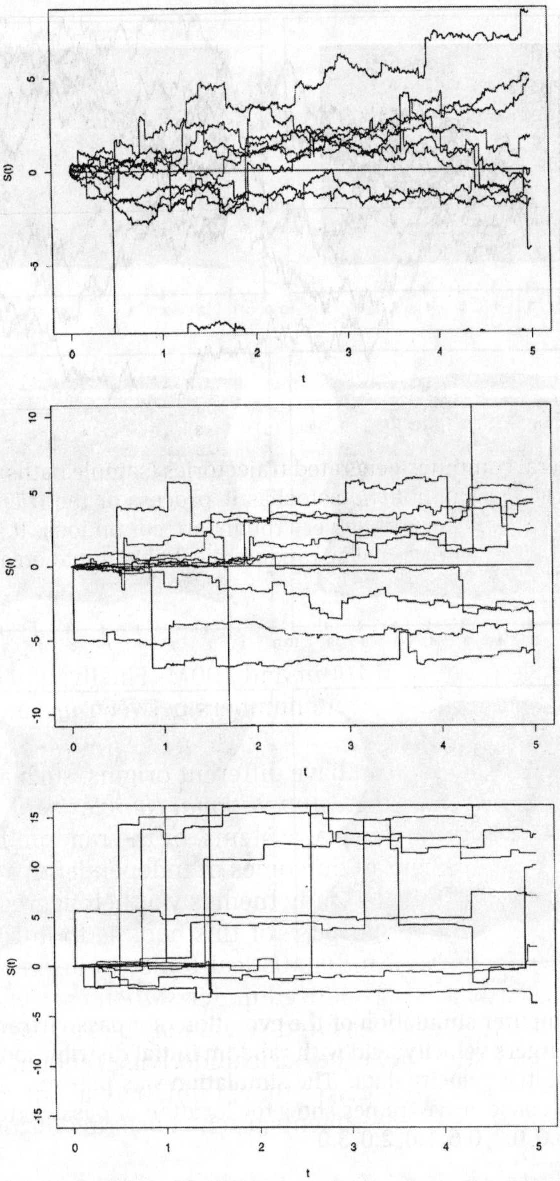

Fig. 1.1.5. Several computer-generated trajectories (sample paths) of random signals called *Lévy stochastic processes* with parameter $\alpha = 1.5$, 1, and 0.75, respectively (from top to bottom). They are often used to model anomalous diffusion processes wherein diffusing particles are also permitted to change their position by jumping. Parameter α indicates the intensity of jumps of different sizes. Parameter value $\alpha = 2$ corresponds to the Wiener process with trajectories that have no jumps. In each figure, the random mechanism that created different trajectories was the same. However, different random mechanisms led to trajectories presented in different figures.

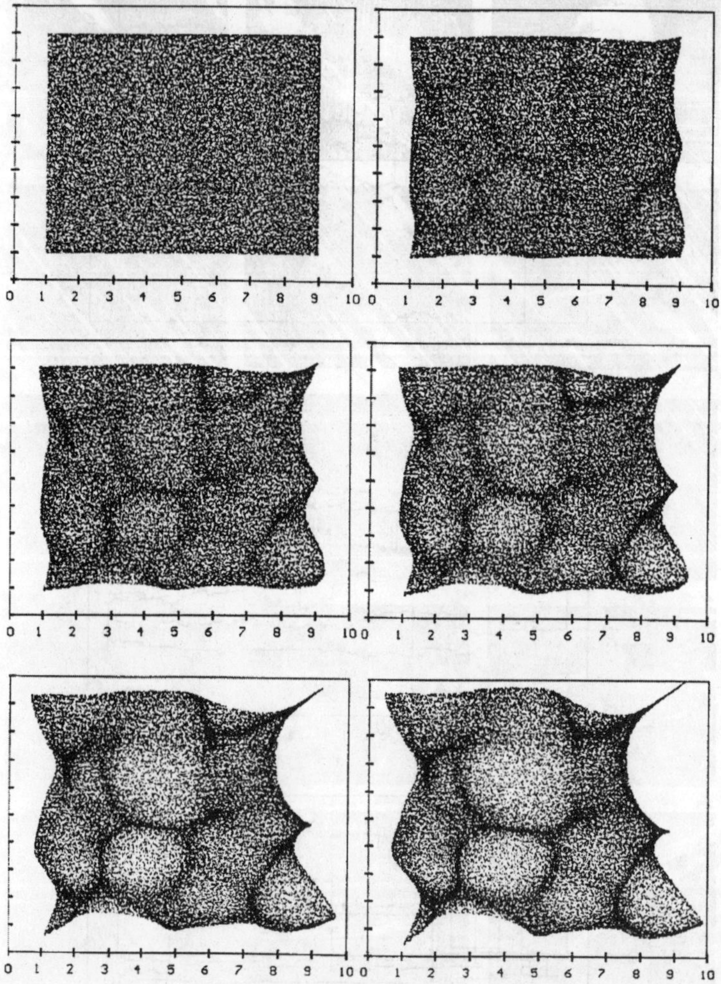

Fig. 1.1.6. Computer simulation of the evolution of a passive tracer density in a turbulent Burgers velocity field with random initial distribution and random "shot-noise" initial velocity data. The simulation was performed for 100,000 particles. The consecutive frames show the location of passive tracer particles at times $t = 0.0, 0.3, 0.6, 1.0, 2.0, 3.0$.

The signals shown in Figures 1.1.4–1.1.5 are, obviously, not stationary and have a diffusive character. However, their increments (differentials) are stationary and, in Chapter 9, they will play an important role in constructing the spectral representation of stationary signals themselves. The signal shown in Figure 1.1.4 can be interpreted as a *trajectory*, or *sample path*, of a *random walker* moving in discrete time steps up or down a certain distance with equal probabilities $\frac{1}{2}$ and $\frac{1}{2}$.

Fig. 1.1.7. Some deterministic signals (in this case, the images) transformed by deterministic systems can appear random. The above picture shows a series of iterated transformations of the original image via a fixed linear 2D mapping (matrix). The number of iterations applied is indicated in the top left corner of each image. The curious behavior of iterations, the original image first dissolving into seeming randomness only to return later to an almost original condition, is related to the so-called *ergodic* behavior. Thus irreverently transformed is Professor Henri Poincaré (1854–1912) of the University of Paris, the pioneer of ergodic theory of stationary phenomena. (From *Scientific American*; reproduced with permission. Copyright 1986 James P. Crutchfield.)

However, in the picture these trajectories are viewed from far away, and in accelerated time, so that both time and space appear continuous.

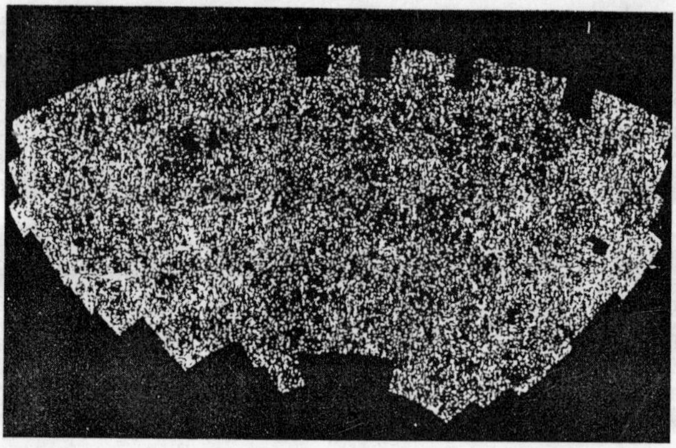

Fig. 1.1.8. A signal (again, an image) representing the large-scale and apparently random distribution of mass in the universe. The data come from the APM galaxy survey and shows more than 2 million galaxies in a section of sky centered on the South Galactic Pole. The so-called *adhesion model* of the large-scale mass distribution in the universe uses the Burgers equation to model the relevant velocity fields.

In certain situations the randomness of the signal is due to uncertainty about initial conditions of the underlying phenomenon which otherwise can be described by perfectly deterministic models such as partial differential equations. A sequence of pictures in Figure 1.1.6 shows the evolution of the system of particles with an initially random (and homogeneous in space) spatial distribution. The particles are then driven by the velocity field $\vec{v}(t,\vec{x}) \in \mathbf{R}^2$ governed by the so-called 2*D Burgers equation*[2]

$$\frac{\partial \vec{v}(t,\vec{x})}{\partial t} + (\vec{v}(t,\vec{x}) \cdot \nabla)\vec{v}(t,\vec{x}) = D\left(\frac{\partial^2 \vec{v}(t,\vec{x})}{\partial x_1} + \frac{\partial^2 \vec{v}(t,\vec{x})}{\partial x_2}\right), \quad (1.1.2)$$

where $\vec{x} = (x_1, x_2)$, the *nabla* operator $\nabla = \frac{\partial}{\partial x_1}\vec{i} + \frac{\partial}{\partial x_2}\vec{j}$, and the positive constant D is the coefficient of diffusivity. The inital velocity field is also assumed to be random.

1.2 Time domain and frequency domain descriptions

A periodic signal with period P (measured, say, in seconds [s]) can be written in the form of an infinite series

$$x(t) = c_0 + \sum_{m=1}^{\infty} c_m \cos(2\pi m f_0 t + \theta_m), \quad (1.2.1)$$

[2] See, e.g., W. A. Woyczyński, *Burgers–KPZ Turbulence–Göttingen Lectures*, Springer-Verlag, Berlin, New York, 1998.

where $f_0 = \frac{1}{P}$ [Hz] is the fundamental frequency of the signal. This expansion, called the *Fourier expansion* of the signal, is the basic tool in the analysis of random signals; it will be reviewed in detail in Chapter 2. The components

$$c_m \cos(2\pi m f_0 t + \theta_m), \quad m = 2, 3, \ldots,$$

are called higher harmonics of the signal with the amplitudes c_m, higher frequencies mf_0, and the corresponding phase shifts θ_m. In the case of zero phase shifts, $\theta_m = 0$, the collection of pairs

$$(mf_0, c_m), \quad m = 1, 2, \ldots,$$

or, equivalently, their graphical representation, is called the *frequency spectrum* of the signal. Note that, for a periodic signal, the spectrum is always concentrated on a discrete set of frequencies, namely, the multiplicities of the fundamental frequency f_0. For example, the signal

$$x(t) = \sum_{m=1}^{12} \frac{1}{m^2} \cos(2\pi m t), \tag{1.2.2}$$

shown in Figure 1.2.1, has the fundamental frequency 1 Hertz (Hz), i.e., one cycle per second, and the frequency spectrum

$$c_m = \begin{cases} m^{-2} & \text{for } m = 1, 2, \ldots, 12, \\ 0 & \text{for } m = 13, 14, \ldots \end{cases} \tag{1.2.3}$$

shown in Figure 1.2.2.

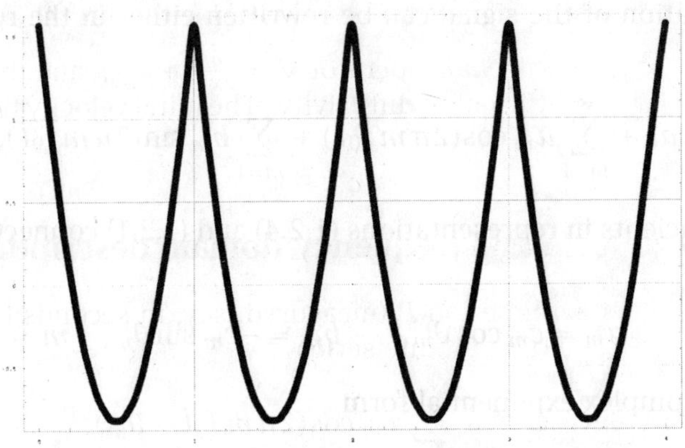

Fig. 1.2.1. Signal $x(t) = \sum_{m=1}^{12} m^{-2} \cos(2\pi m t)$ in its time domain representation.

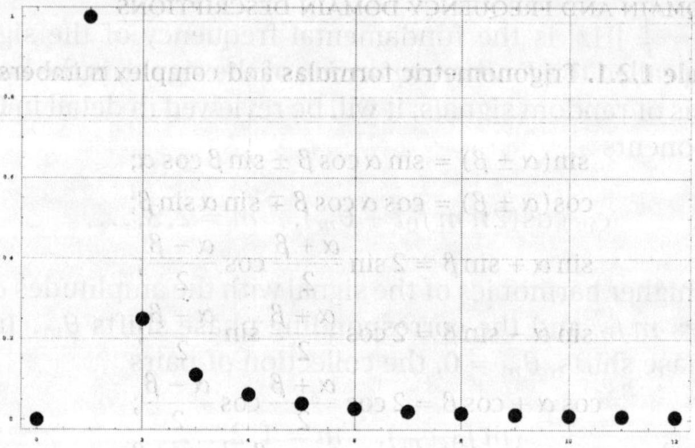

Fig. 1.2.2. Signal $x(t) = \sum_{m=1}^{12} m^{-2}\cos(2\pi mt)$ in its frequency domain (spectral) representation.

If the signal is studied only in a finite time interval $[0, P]$, it can always be treated as a periodic signal with period P since one can extend its definition periodically to the whole time line by copying its waveform from the interval $[0, P]$ to intervals $[P, 2P]$, $[2P, 3P]$, and so on.

Given the familar trigonometric formulas

$$\cos(\alpha + \beta) = \cos\alpha\cos\beta - \sin\alpha\sin\beta$$

and de Moivre's formulas

$$e^{j\alpha} = \cos\alpha + j\sin\alpha, \quad j = \sqrt{-1}, \quad \cos\alpha = \frac{1}{2}(e^{j\alpha} + e^{-j\alpha}),$$

which tie together the trigonometric functions of the real variable α with exponential functions of the imaginary variable $j\alpha$, the spectral representation of the signal can be rewritten either in the real phaseless form

$$x(t) = a_0 + \sum_{m=1}^{\infty} a_m \cos(2\pi m f_0 t) + \sum_{m=1}^{\infty} b_m \sin(2\pi m f_0 t), \qquad (1.2.4)$$

with coefficients in representations (1.2.4) and (1.2.1) connected by the formulas

$$a_0 = c_0, \qquad a_m = c_m\cos\theta_m, \qquad b_m = -c_m\sin\theta_m, \quad m = 1, 2, \ldots,$$

or in the complex exponential form

$$x(t) = \sum_{m=-\infty}^{\infty} z_m e^{j2\pi m f_0 t}, \qquad (1.2.5)$$

Table 1.2.1. Trigonometric formulas and complex numbers.

$$\sin(\alpha \pm \beta) = \sin\alpha\cos\beta \pm \sin\beta\cos\alpha;$$

$$\cos(\alpha \pm \beta) = \cos\alpha\cos\beta \mp \sin\alpha\sin\beta;$$

$$\sin\alpha + \sin\beta = 2\sin\frac{\alpha+\beta}{2}\cos\frac{\alpha-\beta}{2};$$

$$\sin\alpha - \sin\beta = 2\cos\frac{\alpha+\beta}{2}\sin\frac{\alpha-\beta}{2};$$

$$\cos\alpha + \cos\beta = 2\cos\frac{\alpha+\beta}{2}\cos\frac{\alpha-\beta}{2};$$

$$\cos\alpha - \cos\beta = -2\sin\frac{\alpha+\beta}{2}\sin\frac{\alpha-\beta}{2};$$

$$\sin^2\alpha - \sin^2\beta = \cos^2\beta - \cos^2\alpha = \sin(\alpha+\beta)\sin(\alpha-\beta);$$

$$\cos^2\alpha - \sin^2\beta = \cos^2\beta - \sin^2\alpha = \cos(\alpha+\beta)\cos(\alpha-\beta);$$

$$\sin\alpha\cos\beta = \frac{1}{2}[\sin(\alpha+\beta) + \sin(\alpha-\beta)];$$

$$\cos\alpha\cos\beta = \frac{1}{2}[\cos(\alpha+\beta) + \cos(\alpha-\beta)];$$

$$\sin\alpha\sin\beta = \frac{1}{2}[\cos(\alpha-\beta) - \cos(\alpha+\beta)];$$

$$j = \sqrt{-1}, \qquad j^{4m} = 1, \qquad j^{4m+1} = j, \qquad j^{4m+2} = -1, \qquad j^{4m+3} = -j,$$

where m is an integer;

$$z = a + jb, \quad a = \operatorname{Re} z, \quad b = \operatorname{Im} z, \quad z^* = a - jb;$$

$$|z| = \sqrt{a^2 + b^2} = \sqrt{z \cdot z^*};$$

$$\operatorname{Re} z = \frac{z + z*}{2} = |z|\cos\theta, \qquad \operatorname{Im} z = \frac{z - z*}{2j} = |z|\sin\theta,$$

where

$$\theta = \operatorname{Arg} z = \arctan\frac{\operatorname{Im} z}{\operatorname{Re} z}$$

is the *argument* of z.

Table 1.2.2. De Moivre formulas.

$$e^{\beta + j\alpha} = e^{\beta}(\cos\alpha + j\sin\alpha),$$

$$\cos\alpha = \frac{e^{j\alpha} + e^{-j\alpha}}{2}, \qquad \sin\alpha = \frac{e^{j\alpha} - e^{-j\alpha}}{2j},$$

$$(\cos\theta + j\sin\theta)^n = \cos n\theta + j\sin n\theta.$$

with coefficients (amplitudes) in representations (1.2.5) and (1.2.4) connected by the formulas

$$a_0 = z_0, \qquad a_m = z_m + z_{-m}, \qquad b_m = j(z_m - z_{-m}), \quad m = 1, 2, \ldots.$$

For the complex exponential form (1.2.5) to represent a real-valued signal, that is, for a_0, a_m, b_m, given by the above formulas to be real, the condition $z_{-m} = z_m^*$, where the asterisk denotes the complex conjugate, must be satisfied.

Nonperiodic signals can also be analyzed in terms of their spectra, but those spectra are not discrete. We will study them later on.

At the first sight, the above introduction of complex numbers and functions of complex numbers may seem as an unnecessary complication in the analysis of signals. However, as we will see in subsequent chapters, the calculations within the theory of random signals actually become simpler and more transparent if one operates in the complex domain. The book assumes familiarity with elementary properties of trigonometric functions and complex numbers. However, for the reader's peace of mind, and by popular demand of the readers of the preliminary versions of this book, we summarize the basic formulas in this area in the table below and include a few exercises in Section 1.4 to review basic operational procedures on complex numbers.

1.3 Characteristics of signals

Several physical characteristics of signals are of primary interest.

- *The time average of the signal*: For analog, continuous-time signals, the time average is defined by the formula

$$x_{\text{av}} = \lim_{T\to\infty} \frac{1}{T}\int_0^T x(t)\,dt, \tag{1.3.1}$$

and for digital, discrete-time signals which are defined only for the time instants $t = n$, $n = 0, 1, 2, \ldots, N-1$, it is defined by the formula

$$x_{\text{av}} = \frac{1}{N}\sum_{n=0}^{N-1} x(nT). \tag{1.3.2}$$

For periodic signals, it follows from (1.3.1) that

$$x_{\text{av}} = \frac{1}{P}\int_0^P x(t)\,dt, \tag{1.3.3}$$

so that, for signals described by their Fourier expansions, (1.2.1) and (1.2.4)–(1.2.5), the time averages are

$$x_{\text{av}} = c_0 = a_0 = z_0,$$

because the integral of the sine and cosine functions over the full period is 0.

- *Energy of the signal*: For an analog signal $x(t)$, the energy is

$$E_x = \int_0^\infty |x(t)|^2 dt, \tag{1.3.4}$$

and for digital signals,

$$E_x = T \sum_{n=0}^{\infty} |x(nT)|^2. \tag{1.3.5}$$

Remember that, since in what follows it will be convenient to consider complex-valued signals, the above formulas include notation for the square of the modulus of a complex number: $|z|^2 = (\text{Re}\, z)^2 + (\text{Im}\, z)^2 = z \cdot z^*$.

- *Power of the signal*: Again, for an analog signal, the power is

$$\text{PW}_x = \lim_{T\to\infty} \frac{1}{T} \int_0^T |x(t)|^2 dt, \tag{1.3.6}$$

and for a digital signal,

$$\text{PW}_x = \lim_{N\to\infty} \frac{1}{NT} \sum_{n=0}^{N-1} |x(nT)|^2 \cdot T. \tag{1.3.7}$$

As a consequence, for a periodic signal with period P,

$$\text{PW}_x = \frac{1}{P} \int_0^P |x(t)|^2 dt. \tag{1.3.8}$$

Sometimes it is convenient to consider signals defined for all time instants t, $-\infty < t < +\infty$, rather than just for positive t. In such cases, all of the above definitions have to be adjusted in obvious ways, replacing the one-sided integrals and sums by two-sides integrals and sums, and adjusting the averaging constants correspondingly.

1.4 Problems and exercises

1.4.1. Find the real and imaginary parts of $\frac{j+3}{j-3}$; $(1 + j\sqrt{2})^3$; $\frac{1}{2-j}$; $\frac{2-3j}{3j+2}$.

1.4.2. Find the moduli $|z|$ and arguments θ of complex numbers $z = 5$; $z = -2j$; $z = -1 + j$; $z = 3 + 4j$.

1.4.3. Find the real and imaginary components of complex numbers $z = 5e^{j\pi/4}$; $z = -2e^{j(8\pi+1.27)}$; $z = -1e^{j}$; $z = 3e^{je}$.

1.4.4. Show that

$$\frac{5}{(1-j)(2-j)(3-j)} = \frac{j}{2} \quad \text{and} \quad (1-j)^4 = -4.$$

1.4.5. Sketch sets of points in the complex plane (x, y), $z = x + jy$, such that $|z - 1 + j| = 1$; $|z + j| \leq 3$; $\text{Re}(z^* - j) = 2$; $|2z - j| = 4$; $z^2 + (z^*)^2 = 2$.

1.4.6. Using de Moivre's formulas, find $(-2j)^{1/2}$ and $\text{Re}(1 - j\sqrt{3})^{77}$. Are these complex numbers uniquely defined?

1.4.7. Write the signal $x(t) = \sin t + \cos\frac{3t}{3}$ from Figure 1.1.1 in the pure cosine form (1.2.1). Use the fact that sine can be written as a cosine with a phase shift.

1.4.8. Using de Moivre's formulas, write the signal $x(t) = \sin t + \cos\frac{3t}{3}$ from Figure 1.1.1 in the complex exponential form (1.2.5).

1.4.9. Find the time average and power of the signal $x(t) = \sin t + \cos\frac{3t}{3}$ from Figure 1.1.1.

1.4.10. Using de Moivre's formula, derive the complex exponential representation (1.2.5) of the signal $x(t)$ given by the cosine series representation (1.2.1). Then apply this procedure to obtain the complex exponential representation of the signal given by formula (1.2.2) and shown in Figure 1.2.1.

1.4.11. Find the time average and power of the signal $x(t)$ from Figure 1.2.1. Use a symbolic manipulation language such as *Mathematica* or MATLAB if you like.

1.4.12. Verify that for the signal $x(t)$ in (1.2.5) to be real valued, condition $z_{-m} = z_m^*$ has to be satisfied for all integers m.

1.4.13. Using a computing platform such as *Mathematica*, MAPLE, or MATLAB, produce plots of the signals

$$x_n(t) = \frac{\pi}{4} + \sum_{m=1}^{M}\left[\frac{(-1)^m - 1}{\pi m^2}\cos mt - \frac{(-1)^m}{m}\sin mt\right]$$

for $M = 0, 1, 2, 3, \ldots, 9$ and $-2\pi < t < 2\pi$. Then produce their plots in the frequency domain representation. Calculate their power (again, using *Mathematica*, MAPLE, or MATLAB if you wish). Write down your observations. What is likely to happen with the plots of these signals as we take more and more terms of the above series, that is, as $M \to \infty$?

1.4.14. Use the analog-to-digital conversion formula (1.1.1) to digitize signals from Problem 1.4.13 for a variety of sampling periods and resolutions. Plot the results.

1.4.15. Use your computing platform to produce a discrete-time signal consisting of a string of random numbers uniformly distributed on the interal [0, 1]. For example, in *Mathematica*, the command

```
Table[Random[], {20}]
```

may produce the following string of 20 random numbers between 0 and 1:

```
{0.175245, 0.552172, 0.471142, 0.910891, 0.219577,
 0.198173, 0.667358, 0.226071, 0.151935, 0.42048,
 0.264864, 0.330096, 0.346093, 0.673217, 0.409135,
 0.265374, 0.732021, 0.887106, 0.697428, 0.7723}
```

Use the "random numbers" string as additive noise to produce random versions of the digitized signals from Problem 1.4.14. Follow the example described in Figure 1.1.3. Experiment with different string lengths and various noise amplitudes. Then center the noise around zero and repeat your experiments.

2

Spectral Representation of Deterministic Signals: Fourier Series and Transforms

In this chapter we will take a closer look at the spectral, or frequency domain, representation of deterministic (nonrandom) signals which was already mentioned in Chapter 1. The tools introduced below, usually called *Fourier* or *harmonic analysis*, will play a fundamental role later on in our study of random signals. Almost all of the calculations will be conducted in the complex form. Compared with working in the real domain, manipulation of formulas written in the complex form turns out to be simpler and all the tedium of remembering various trigonometric formulas is avoided. All of the results written in the complex form can be translated quickly into results for real trigonometric series expressed in terms of sines and cosines via de Moivre's formula $e^{jt} = \cos t + j \sin t$, familiar from Chapter 1.

2.1 Complex Fourier series for periodic signals

A complex-valued signal $x(t)$ that is periodic with period P (say, seconds) can be written in the form of an infinite complex Fourier series

$$x(t) = \sum_{m=-\infty}^{\infty} z_m e^{j2\pi m f_0 t} = \sum_{m=-\infty}^{\infty} z_m e^{jm\omega_0 t}, \qquad (2.1.1)$$

where $f_0 = \frac{1}{P}$ is the *fundamental frequency* of the signal (measured in Hz $= \frac{1}{s}$), and $\omega_0 = 2\pi f_0$ is called the fundamental *angular velocity* (measured in radians/s). The complex number z_m, where m can take values $\ldots, -2, -1, 0, 1, 2, \ldots$, is called the mth Fourier coefficient of signal $x(t)$.

In this text, we will carry out our calculations exclusively in terms of the fundamental frequency f_0, although one can find in the printed and software signal processing literature sources where all the work is done in terms of ω_0. It is an arbitrary choice, and transition from one system to the other is easily accomplished by adjusting various constants appearing in the formulas.

The infinite Fourier series representation (2.1.1) is unique in the sense that two different signals will have two different sequences of Fourier coefficients. The uniqueness is a result of the fundamental property of complex exponentials

$$e_m(t) := e^{j2\pi m f_0 t}, \quad m = \ldots, -2, -1, 0, 1, 2, \ldots, \tag{2.1.2}$$

called *orthonormality*:

The scalar product (sometimes also called inner, or dot, product) of two complex exponentials e_n and e_m is 0 if the exponentials are different, and it is 1 if they are the same. Indeed,

$$\begin{aligned}\langle e_n, e_m\rangle &= \frac{1}{P}\int_0^P e_n(t)e_m^*(t)\,dt \\ &= \frac{1}{P}\int_0^P e^{j2\pi(n-m)f_0 t}\,dt = \begin{cases}0 & \text{if } n \neq m; \\ 1 & \text{if } n = m.\end{cases}\end{aligned} \tag{2.1.3}$$

Recall that, for a complex number $z = a + jb = |z|e^{j\theta}$ with real component a and imaginary component b, the complex conjugate $z^* = a - jb = |z|e^{-j\theta}$. Sometimes it is convenient to describe the orthonormality using the so-called *Kronecker delta* notation:

$$\delta_{mn} = \begin{cases}0 & \text{if } n \neq m; \\ 1 & \text{if } n = m.\end{cases}$$

Then, simply,

$$\langle e_m, e_n\rangle = \delta_{mn}.$$

Using the orthonormality property we can directly evaluate the coefficients z_m in the Fourier series (2.1.1) of signal $x(t)$ by formally calculating the scalar product of $x(t)$ and $e_m(t)$:

$$\begin{aligned}\langle x, e_m\rangle &= \frac{1}{P}\int_0^P \left(\sum_{n=-\infty}^{\infty} z_n e_n(t)\right)\cdot e_m^*(t)\,dt \\ &= \sum_{n=-\infty}^{\infty} z_n \frac{1}{P}\int_0^P e_n(t)e_m^*(t)\,dt = z_m,\end{aligned} \tag{2.1.4}$$

so that we get an explicit formula for the Fourier coefficent of signal $x(t)$,

$$z_m = \langle x, e_m \rangle = \frac{1}{P}\int_0^P x(t)e^{-j2\pi m f_0 t}\,dt. \tag{2.1.5}$$

Thus the basic Fourier expansion (2.1.1) can now be rewritten in the form of a formal identity

$$x(t) = \sum_{n=-\infty}^{\infty} \langle x, e_n \rangle e_n(t). \tag{2.1.6}$$

It is worthwhile to observe that the above calculations on infinite series and interchanges of the order of integration and infinite summations were purely formal, that is, the soundness of the limit procedures was not rigorously established. The missing steps can be found in the mathematical literature devoted to Fourier analysis.[3] For our purposes, it suffices to say that if a periodic signal $x(t)$ has finite power

$$\mathrm{PW}_x = \frac{1}{P}\int_0^P |x(t)|^2 dt < \infty, \tag{2.1.7}$$

and the concept of convergence of the functional infinite series (2.1.1) is defined in the right way, then all of the above formal manipulations can be rigorously justified. We will return to this issue at the end of this section. In what follows, we will usually consider signals with finite power.

Real-valued signals. Signal $x(t)$ is real-valued if and only if the coefficients z_m satisfy the obvious algebraic condition

$$z_{-m} = z_m^*, \tag{2.1.8}$$

in which case cancellation of the imaginary parts in the Fourier series (2.1.1) occurs. Indeed, under assumption (2.1.8),

$$z_m = |z_m|e^{j\theta_m}, \quad \theta_{-m} = -\theta_m, \tag{2.1.9}$$

and since

$$\frac{e^{j\alpha} + e^{-j\alpha}}{2} = \cos\alpha,$$

we get

$$x(t) = c_0 + \sum_{m=1}^{\infty} c_m \cos(2\pi m f_0 t + \theta_m), \tag{2.1.10}$$

where

[3] See, e.g., A. Zygmund, *Trigonometric Series*, Cambridge University Press, Cambridge, UK, 1959.

$$c_0 = z_0 \quad \text{and} \quad c_m = 2|z_m|, \quad m = 1, 2, \ldots. \tag{2.1.11}$$

The power PW_x of a periodic signal $x(t)$ can also be directly calculated from its Fourier coefficient z_m. Indeed, again calculating formally, we obtain that

$$\begin{aligned} PW_x &= \frac{1}{P}\int_0^P |x(t)|^2 dt = \frac{1}{P}\int_0^P x(t)x^*(t)dt \\ &= \frac{1}{P}\int_0^P \left(\sum_{k=-\infty}^{\infty} z_k e_k(t)\right) \cdot \left(\sum_{m=-\infty}^{\infty} z_m e_m(t)\right)^* dt \\ &= \sum_{k=-\infty}^{\infty}\sum_{m=-\infty}^{\infty} z_k z_m^* \frac{1}{P}\int_0^P e_k(t)e_m^*(t)dt = \sum_{m=-\infty}^{\infty} z_m z_m^*, \end{aligned}$$

in view of the orthonormality (2.1.3) of the complex exponentials. The multiplication of the two infinite series was carried out term by term. The resulting relationship,

$$PW_x = \frac{1}{P}\int_0^P |x(t)|^2 dt = \sum_{m=-\infty}^{\infty} |z_m|^2, \tag{2.1.12}$$

is known as the *Parseval formula*. A similar calculation for the scalar product $\frac{1}{P}\int_0^P x(t)y^*(t)dt$ of two different periodic signals, $x(t)$ and $y(t)$, gives an *extended Parseval formula* listed in Table 2.1.1.

Analogy between the orthonormal basis of vectors in the 3D space $\mathbf{R}^3$ and the complex exponentials. It is useful to think about the complex exponentials $e_m(t) = e^{2\pi jmf_0t}$, $m = \ldots, -1, 0, 1, \ldots$, as an infinite-dimensional version of the orthonormal basics vectors in $\mathbf{R}^3$. In this mental picture the periodic signal $x(t)$ is now thought of as an infinite-dimensional "vector" uniquely expandable into an infinite linear combination of the complex exponentials in the same way a 3D vector is uniquely expandable into a finite linear combination of the three unit coordinate vectors. Table 2.1.1 describes this analogy more fully. Note that the Parseval formula can now be seen just as an infinite-dimensional extension of the familiar Pythagorean theorem.

Recall that a signal is called *even* if it is symmmetric undcr the change of the direction of time, i.e., if $x(t) = x(-t)$; it is called *odd* if it is antisymmetric under the change of the direction of time, i.e., if $x(t) = -x(-t)$. The real Fourier expansion of an even real-valued signal $x(t) = x(-t)$ will contain only cosine functions, and the real Fourier expansion of an odd real-valued signal $x(t) = -x(-t)$ will contain only sine functions. This phenomenon will be illustrated in the following examples. Of course, if one is only interested in the signal $x(t)$ for positive times $t > 0$, then one can arbitrarily extend the signal's values to

Table 2.1.1. Analogy between orthonormal expansions in 3D and in the space of periodic signals.

Objects	
3D vectors	Signals with finite power
$\vec{x} = (x_1, x_2, x_3)$	$x(t) = \sum_{m=-\infty}^{\infty} z_m e_m(t)$, $\mathrm{PW}_x < \infty$
$\vec{y} = (y_1, y_2, y_3)$	$y(t) = \sum_{m=-\infty}^{\infty} w_m e_m(t)$, $\mathrm{PW}_y < \infty$
Bases	
Unit coordinate vectors	Complex exponentials
	$\vdots$
$\vec{e_1} = (1, 0, 0)$	$e_1(t) = e^{j2\pi f_0 t}$
$\vec{e_2} = (0, 1, 0)$	$e_2(t) = e^{j2\pi(2f_0)t}$
$\vec{e_3} = (0, 0, 1)$	$e_3(t) = e^{j2\pi(3f_0)t}$
	$\vdots$
Scalar products	
$\langle \vec{x}, \vec{y} \rangle = \sum_{i=1}^{3} x_i y_i$	$\langle x(t), y(t) \rangle = \frac{1}{P}\int_0^P x(t) y^*(t) dt$
Orthonormality	
$\langle \vec{e_m}, \vec{e_m} \rangle = \delta_{mn}$	$\langle e_m(t), e_n(t) \rangle = \delta_{mn}$
Expansions	
Basis	Fourier
$\vec{x} = \sum_{m=1}^{3} \langle \vec{x}, \vec{e_m} \rangle \vec{e_m}$	$x(t) = \sum_{i=-\infty}^{\infty} \langle x, e_m \rangle e_m(t)$
Formulas	
Pythagoras'	Parseval's
$\lVert\vec{x}\rVert^2 = \sum_{m=1}^{3} x_m^2$	$\mathrm{PW}_x = \frac{1}{P}\int_0^P \lvert x(t)\rvert^2 dt = \sum_{m=-\infty}^{\infty} \lvert z_m\rvert^2$
Scalar product	Extended Parseval's
$\langle \vec{x}, \vec{y} \rangle = \sum_{m=1}^{3} x_m y_m$	$\frac{1}{P}\int_0^P x(t) y^*(t) dt = \sum_{m=-\infty}^{\infty} z_m w_m^*$

the negative timeline to form either an odd or an even signal, and thus obtain either its pure sine or its pure cosine expansion.

Example 2.1.1 (pure cosine expansion of an even rectangular waveform). Consider a rectangular waveform with period P and amplitude

$a > 0$, defined by the formula

$$x(t) = \begin{cases} a & \text{for } 0 \le t < \frac{P}{4}; \\ 0 & \text{for } \frac{P}{4} \le t < \frac{3P}{4}; \\ a & \text{for } \frac{3P}{4} \le t < P. \end{cases}$$

The signal is pictured in Figure 2.1.1 for particular values $P = 1$ and $a = 1$.

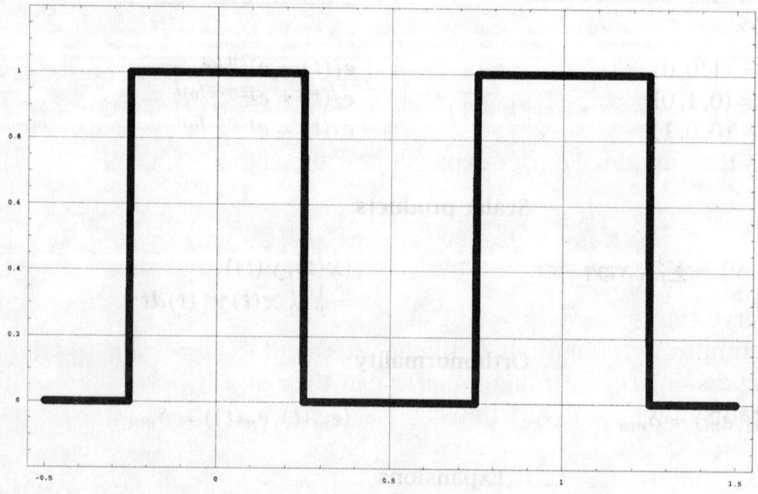

Fig. 2.1.1. An even rectangular waveform signal from Example 2.1.1. The period $P = 1$ and the amplitude $a = 1$.

Calculation of coefficients z_m in the expansion of the signal $x(t)$ into a complex Fourier series is here straightforward: For $m = 0$,

$$z_0 = \frac{1}{P}\int_0^P x(t)e^{-j2\pi 0t/P}\,dt = \frac{a}{P}\left(\frac{P}{4} - 0 + P - \frac{3P}{4}\right) = \frac{a}{2}.$$

In the case $m \neq 0$,

$$\begin{aligned} z_m &= \frac{1}{P}\int_0^P x(t)e^{-j2\pi mt/P}\,dt \\ &= \frac{a}{P}\left(\int_0^{P/4} e^{-j2\pi mt/P}\,dt + \int_{3P/4}^{P} e^{-j2\pi mt/P}\,dt\right) \\ &= \frac{a}{P}\left(\frac{P}{-j2\pi m}e^{-j2\pi mt/P}\Big|_0^{P/4} + \frac{P}{-j2\pi m}e^{-j2\pi mt/P}\Big|_{3P/4}^{P}\right) \\ &= \frac{a}{-j2\pi m}\left(e^{-j(\pi/2)m} - 1 - e^{-j(3\pi/2)m} + 1\right) \end{aligned}$$

$$= -\frac{a}{\pi m} e^{-j(2\pi/2)m} \left(\frac{e^{j(\pi/2)m} - e^{-j(\pi/2)m}}{2j} \right)$$
$$= -\frac{a}{\pi m} \cos \pi m \sin \frac{\pi}{2} m = -\frac{a}{\pi m} (-1)^m \sin \frac{\pi}{2} m.$$

If $m = 2k$, then $\sin \frac{\pi}{2} m = 0$, and if $m = 2k+1$, $k = 0, \pm 1, \pm 2, \ldots$, then $\sin \frac{\pi}{2} m = (-1)^k$, which gives, for $k = \pm 1, \pm 2, \ldots$,

$$z_{2k} = 0,$$

and

$$z_{2k+1} = \frac{-a}{\pi(2k+1)} (-1)^{2k+1} (-1)^k = \frac{(-1)^k a}{\pi(2k+1)}.$$

Thus the complex Fourier expansion of the signal $x(t)$ is

$$x(t) = \frac{a}{2} + \frac{a}{\pi} \sum_{k=-\infty}^{\infty} \frac{(-1)^k}{2k+1} e^{j2\pi(2k+1)t/P}.$$

Observe that for any $m = \ldots, -1, 0, 1, \ldots$, we have $z_m = z_{-m}$. Pairing up complex exponentials with the exponents of opposite signs, and using de Moivre's formula, we arrive at the real Fourier expansion that contains only cosine functions:

$$x(t) = \frac{a}{2} + \frac{a}{\pi} \left(2 \cos \frac{2\pi t}{P} - \frac{2}{3} \cos \frac{2\pi 3t}{P}) + \cdots \right).$$

Example 2.1.2 (pure sine expansion of an odd rectangular waveform). Consider a periodic rectangular waveform of period P which is defined by the formula

$$x(t) = \begin{cases} a & \text{for } 0 \le t < \frac{P}{4}; \\ 0 & \text{for } \frac{P}{4} \le t < \frac{3P}{4}; \\ -a & \text{for } \frac{3P}{4} \le t < P. \end{cases}$$

The signal is pictured in Figure 2.1.2 for particular values $P = 1$ and $a = 1$.

For $m = 0$,

$$z_0 = \frac{1}{P} \int_0^P x(t)dt = 0,$$

and, for $m \ne 0$,

$$z_m = \frac{a}{P} \left(\int_0^{P/4} e^{-j2\pi mt/P} dt - \int_{3P/4}^{P} e^{-j2\pi mt/P} dt \right)$$
$$= \frac{a}{j2\pi m} (e^{-j(\pi/2)m} - 1 - 1 + e^{-j(3\pi/2)m})$$

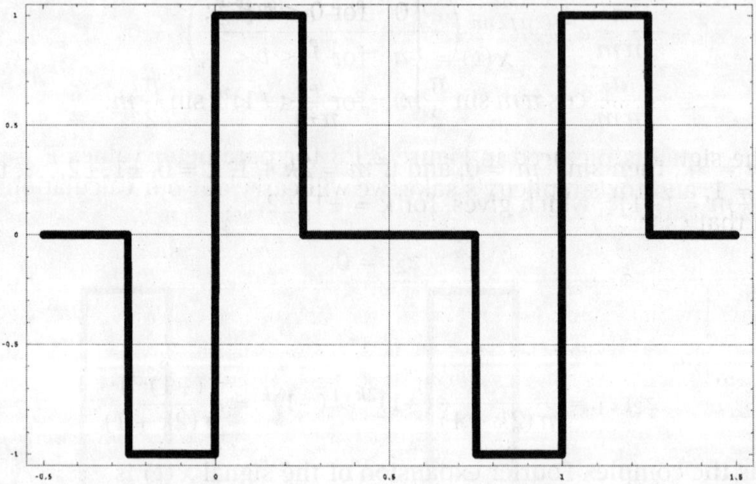

Fig. 2.1.2. An odd rectangular waveform signal from Example 2.1.2. The period $P = 1$ and the amplitude $a = 1$.

$$= -\frac{aj}{2\pi m}[e^{-j(2\pi/2)m}(e^{j(\pi/2)m} + e^{-j(\pi/2)m}) - 2]$$

$$= -\frac{aj}{\pi m}\left(\cos \pi m \cdot \cos \frac{\pi}{2} m - 1\right).$$

Since $\cos \pi m = (-1)^m$, and since $\cos(\frac{\pi}{2})m = 0$ if m is odd and $= (-1)^k$ when $m = 2k$ is even, we get that

$$z_m = \begin{cases} 0 & \text{for odd } m = 2k+1; \\ \frac{aj[(-1)^k - 1]}{2\pi k} & \text{for even } m = 2k. \end{cases}$$

Thus the complex Fourier series of the signal $x(t)$ is of the form

$$x(t) = \frac{a}{\pi} \sum_{k=-\infty}^{\infty} \frac{j[(-1)^k - 1]}{2k} e^{j2\pi 2kt/P}.$$

Observe that in this case, for any $m = \ldots, -1, 0, 1, \ldots$, we have $z_m = -z_{-m}$, so pairing up the exponentials with opposite signs in the exponents and using de Moivre's formula, we get a real Fourier series expansion for $x(t)$ that contains only sine functions:

$$x(t) = \frac{a}{\pi}\left[2 \sin\left(\frac{4\pi t}{P}\right) + \frac{2}{3} \sin\left(\frac{12\pi t}{P}\right) + \cdots\right].$$

Example 2.1.3 (a general expansion for a rectangular waveform which is neither odd nor even). Consider a periodic rectangular waveform of period P which is defined by the formula

$$x(t) = \begin{cases} 0 & \text{for } 0 \le t < \frac{P}{4}; \\ a & \text{for } \frac{P}{4} \le t < \frac{P}{2}; \\ 0 & \text{for } \frac{P}{2} \le t < P. \end{cases}$$

The signal is pictured in Figure 2.1.3 for parameter values $P = 1$ and $a = 1$, and for simplicity's sake, we will carry out our calculations only in that case.

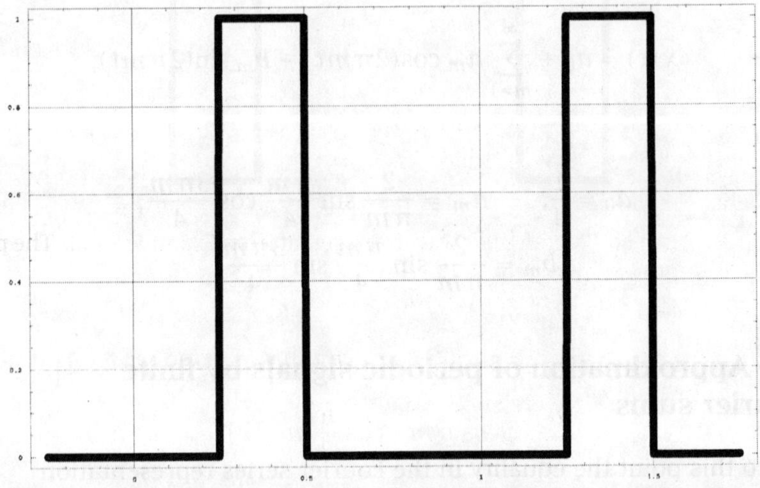

Fig. 2.1.3. A neither odd nor even rectangular waveform signal from Example 2.1.3. The period $P = 1$, and the amplitude $a = 1$.

For $m = 0$,

$$z_0 = \int_{1/4}^{1/2} = \frac{1}{4}.$$

For $m \neq 0$,

$$z_m = |z_m|e^{i\theta m} = \int_{1/4}^{1/2} e^{-j2\pi mt}\,dt = \frac{1}{-j2\pi m}[e^{-j2\pi m/2} - e^{-j2\pi m/4}]$$

$$= \frac{1}{\pi m}e^{-j3\pi m/4}\left(\frac{e^{j\pi m/4} - e^{-j\pi m/4}}{2j}\right) = \frac{1}{\pi m}\sin\left(\frac{\pi}{4}m\right)e^{-j3\pi m/4}.$$

Thus

$$|z_m| = \frac{1}{\pi m}\sin\left(\frac{\pi m}{4}\right) \quad \text{and} \quad \theta_m = -\frac{j3\pi m}{4},$$

and the complex Fourier series for $x(t)$ is

$$x(t) = \frac{1}{4} + \sum_{m=-\infty, m\neq 0}^{\infty} \frac{1}{\pi m}\sin\left(\frac{\pi m}{4}\right)e^{-j3\pi m/4}e^{j2\pi mt}.$$

Again, pairing up the complex exponentials with opposite signs in the exponents, we obtain the real expansion in terms of the cosines, but this time with phase shifts that depend on m:

$$x(t) = \frac{1}{4} + \sum_{m=1}^{\infty} \frac{2}{\pi m} \sin\left(\frac{\pi m}{4}\right) \cos\left(2\pi m t - \frac{3\pi m}{4}\right),$$

which, using the trigonometric formula $\cos(\alpha + \beta) = \cos\alpha\cos\beta - \sin\alpha\sin\beta$, can be written as a general real Fourier series

$$x(t) = a_0 + \sum_{m=1}^{\infty} a_m \cos(2\pi m t) + b_m \sin(2\pi m t),$$

with

$$a_0 = \frac{1}{4}, \qquad a_m = \frac{2}{\pi m} \sin\frac{\pi m}{4} \cos\frac{3\pi m}{4},$$
$$b_m = \frac{2}{\pi m} \sin\frac{\pi m}{4} \sin\frac{3\pi m}{4}.$$

2.2 Approximation of periodic signals by finite Fourier sums

Up to this point the equality in the Fourier series representation

$$x(t) = \sum_{m=-\infty}^{\infty} \langle x, e_m \rangle e_m(t)$$

for periodic signals, or its real version in terms of sine and/or cosine functions, was understood only formally. But, of course, the usefulness of such an expansion will depend on whether we can show that the signal $x(t)$ can be well approximated by a finite cutoff of the infinite Fourier series, that is, on whether we can prove that

$$x(t) \approx s_M(t) := \sum_{m=-M}^{M} \langle x, e_m \rangle e_m(t) \tag{2.2.1}$$

for M large enough, with the error in the above approximate equality $\approx$ rigorously estimated.

One can pursue here several options:

Approximation in power: Mean-square error. If the error of approximation is measured as the power of the difference between the signal $x(t)$ and the finite Fourier sum $s_M(t)$ in (2.2.1), then the calculation is relatively simple and the error is often called the *mean-square error*. Indeed, using the Parseval formula, we find that

$$\begin{aligned}
\mathrm{PW}_{x-s_M} &= \frac{1}{P}\int_0^P |x(t)-s_M(t)|^2\,dt \\
&= \frac{1}{P}\int_0^P \left|\sum_{m=-\infty}^{\infty}\langle x, e_m\rangle e_m(t) - s_M(t)\right|^2 dt \\
&= \frac{1}{P}\int_0^P \left|\sum_{|m|>M}\langle x, e_m\rangle e_m(t)\right|^2 dt = \sum_{|m|>M} |\langle x, e_m\rangle|^2,
\end{aligned}$$

which converges to 0, as $M \to \infty$, because we assumed that the power of the signal is finite:

$$\mathrm{PW}_x = \sum_{m=-\infty}^{\infty} |\langle x, e_m\rangle|^2 < \infty.$$

Note that the unspoken assumption here is that the orthonormal system $e_n(t)$, $n = 0, \pm1, \pm2, \ldots$, is rich enough to make the Fourier representation possible for any finite power signal. This assumption, often called *completeness* of the above orthonormal system, can actually be rigorously proven.

Approximation at each time instant t separately. This type of approximation is often called the *pointwise approximation* and the goal is to verify that, for each time instant t,

$$\lim_{M\to\infty} s_M(t) = x(t). \tag{2.2.2}$$

Here the situation is delicate, as examples at the end of this section will show, and the assumption that signal $x(t)$ has finite power is not sufficient to guarantee the pointwise approximation. Neither is a stronger assumption that the signal is continuous. However,

---

if the signal is continuous and has a bounded continuous derivative, except, possibly, at a finite number of points, then the pointwise approximation (2.2.2) *holds true.*

Uniform approximation in time t. If one wants to control the error of approximation simultaneously (uniformly) for all times t, then more stringent assumptions on the signal are necessary. Namely, we have the following theorem:[4]

[4] Proofs of these two mathematical theorems and other results quoted in this section can be found in, e.g., T. W. Körner, *Fourier Analysis*, Cambridge University Press, Cambridge, UK, 1988.

If the signal is continuous everywhere and has a bounded continuous derivative except at a finite number of points, then

$$\max_{0\le t\le P} |x(t) - s_M(t)| \to 0 \quad \textit{as } M \to \infty. \tag{2.2.3}$$

Note that the above statements do not resolve the question of what happens with the finite Fourier sums at discontinuity points of a signal, like those encountered in the rectangular waveforms in Examples 2.1.1–2.1.3. It turns out that under the assumptions of the above-quoted theorems, the points of discontinuity of the signal $x(t)$ are necessarily jumps, that is the left and right limits

$$x(t_-) = \lim_{s\uparrow t} x(s) \quad \text{and} \quad x(t_+) = \lim_{s\downarrow t} x(s) \tag{2.2.4}$$

exist, and the finite Fourier sums $s_M(x)$ of $x(t)$ converge, as $M \to \infty$, to the average value of the signal at the jump:

$$\lim_{M\to\infty} s_M(t) = \frac{x(t_-) + x(t_+)}{2}. \tag{2.2.5}$$

Example 2.2.1. For the signal $x(t)$ in Example 2.1.1, the first three nonzero terms of its cosine expansion were

$$x(t) = \frac{a}{2} + \frac{a}{\pi}\left(2\cos\left(2\pi\frac{t}{P}\right) - \frac{2}{3}\cos\left(2\pi\frac{3t}{P}\right) + \cdots\right).$$

Hence, in the case of period $P = 1$ and amplitude $a = 1$, the first four approximating sums are as follows:

$$s_0(t) = \frac{1}{2}, \qquad s_1(t) = \frac{1}{2} + \frac{2}{\pi}\cos 2\pi t,$$

$$s_2(t) = \frac{1}{2} + \frac{2}{\pi}\cos 2\pi t, \qquad s_3(t) = \frac{1}{2} + \frac{2}{\pi}\cos 2\pi t - \frac{2}{3\pi}\cos 6\pi t.$$

The graphs of $s_1(t)$ and $s_3(t)$ are compared with the original signal $x(t)$ in Figures 2.2.1–2.2.2. Note the behavior of the Fourier sums at the signal's discontinuities where the Fourier sums converge to the average value of the signal on both sides of the jump according to formula (2.2.5).

Remark. A word of warning is appropriate here. Abandoning the assumptions in the above two theorems leads very quickly to difficulties with approximating the signal by its Fourier series. For example, there are continuous signals for which, at some time instants, their finite

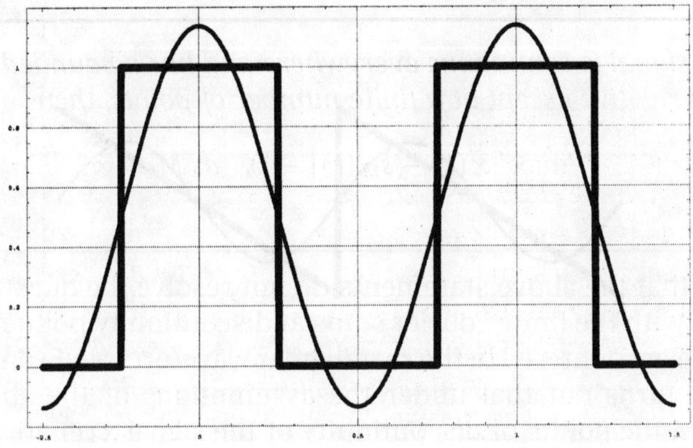

Fig. 2.2.1. Graph of the Fourier sum $s_1(t)$ for the rectangular waveform signal $x(t)$ from Example 2.1.1, plotted against the original signal $x(t)$.

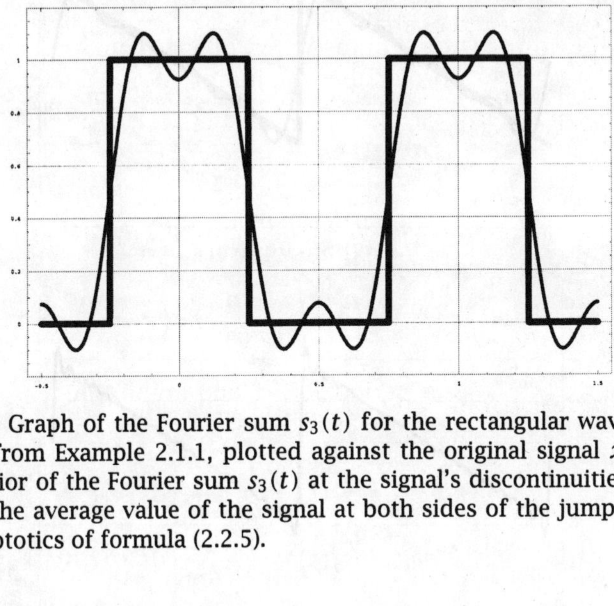

Fig. 2.2.2. Graph of the Fourier sum $s_3(t)$ for the rectangular waveform signal $x(t)$ from Example 2.1.1, plotted against the original signal $x(t)$. Note the behavior of the Fourier sum $s_3(t)$ at the signal's discontinuities, where it matches the average value of the signal at both sides of the jump, reflecting the asymptotics of formula (2.2.5).

Fourier sums diverge to infinity. However, even for them, one can guarantee that the averages of consecutive Fourier sums converge to the signal for each t:

$$\frac{s_0(t) + s_1(t) + \cdots + s_M(t)}{M+1} \to x(t) \quad \text{as } M \to \infty.$$

The expression on the left-hand side of the above formula is called the Mth *Césaro average* of the Fourier series. If one only assumes that the signal $x(t)$ is integrable, that is $\int_0^P |x(t)|dt < \infty$, which is the minimum assumption assuring that the Fourier coefficients $z_m = \langle x, e_m \rangle$ make

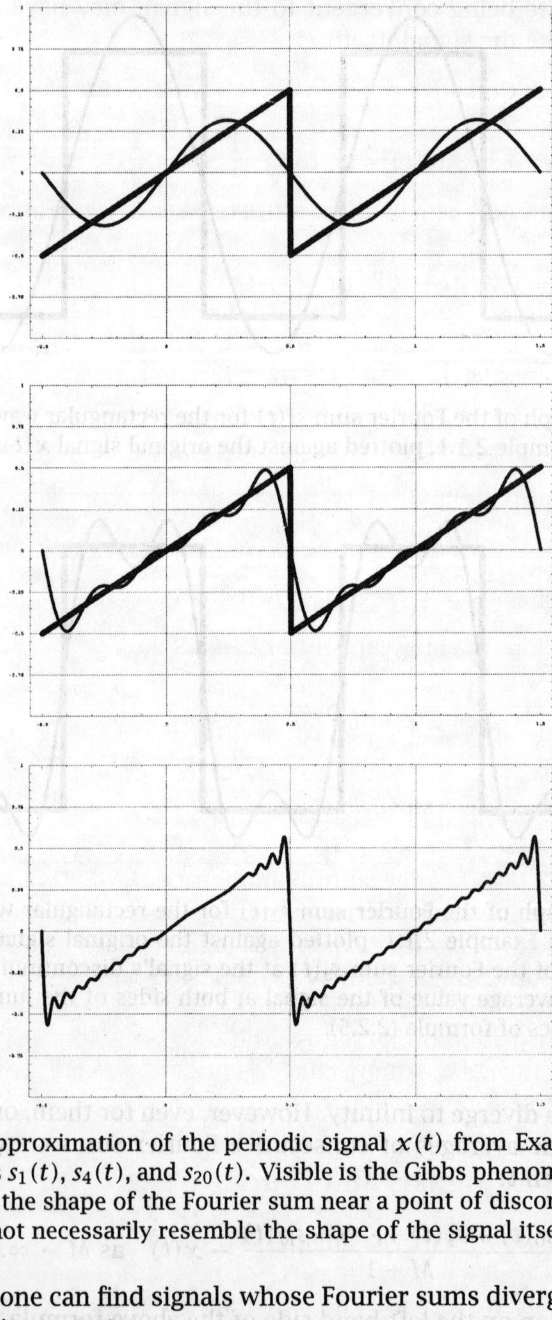

Fig. 2.2.3. Approximation of the periodic signal $x(t)$ from Example 2.2.2 by Fourier sums $s_1(t)$, $s_4(t)$, and $s_{20}(t)$. Visible is the Gibbs phenomenon demonstrating that the shape of the Fourier sum near a point of discontinuity of the signal does not necessarily resemble the shape of the signal itself.

sense, then one can find signals whose Fourier sums diverge to infinity, for all time instants t.

The Gibbs phenomenon. Another observation is that the finite Fourier sums of a signal satisfying the assumptions of the above-quoted statements, despite being convergent to the signal, may have shapes that are very unlike the signal itself.

Example 2.2.2. Consider the signal $x(t)$, with period $P = 1$, defined by the formula

$$x(t) = t \quad \text{for} \quad -\frac{1}{2} \le t < \frac{1}{2}.$$

Clearly, it is an odd signal, so $z_0 = 0$. For $m \neq 0$, integrating by parts,

$$\begin{aligned} z_m &= \int_{-1/2}^{1/2} te^{-j2\pi mt}\,dt \\ &= t\frac{-1}{j2\pi m}e^{-j2\pi mt}\Big|_{-1/2}^{1/2} - \frac{-1}{j2\pi m}\int_{-1/2}^{1/2} e^{-j2\pi mt}\,dt \\ &= -\frac{1}{j2\pi m}(-1)^m \end{aligned}$$

because the last integral is zero. The complex Fourier expansion of $x(t)$ is

$$x(t) = \sum_{m=-\infty, m\neq 0}^{\infty} -\frac{1}{j2\pi m}(-1)^m e^{j2\pi mt},$$

which yields a pure sine real Fourier expansion

$$\begin{aligned} x(t) &= \sum_{m=1}^{\infty}\left(-\frac{1}{j2\pi m}(-1)^m e^{j2\pi mt} + -\frac{1}{j2\pi(-m)}(-1)^{-m}e^{j2\pi(-m)t}\right) \\ &= \sum_{m=1}^{\infty}\frac{(-1)^{m+1}}{\pi m}\sin(2\pi mt). \end{aligned}$$

Figure 2.2.3 shows the approximation of the periodic signal $x(t)$ from Example 2.2.2 by Fourier sums $s_1(t)$, $s_4(t)$, and $s_{20}(t)$. Visible is the so-called *Gibbs phenomenon* demonstrating that the shape of the Fourier sum near a point of discontinuity of the signal does not necessarily resemble the shape of the signal itself. Yet, as the order M of the approximation increases, the oscillations move closer to the jump so that the mean-square convergence of finite Fourier sums to the signal $x(t)$ still obtains.

2.3 Aperiodic signals and Fourier transforms

Periodic signals with increasing period: From Fourier series to Fourier transform. Consider a signal $x_P(t)$ of period P and fundamental frequency $f_0 = \frac{1}{P}$. We already know that such signals can be represented by this Fourier series

$$x_P(t) = \sum_{m=-\infty}^{\infty}\left[\frac{1}{P}\int_{-P/2}^{P/2} x(s)e^{-j2\pi mf_0 s}\,ds\right]\cdot e^{j2\pi mf_0 t}. \tag{2.3.1}$$

Notice that, for the purposes of this section, we have written the formula for the Fourier coefficients of $x_P(t)$ as an integral over a symmetric interval $(-\frac{P}{2}, \frac{P}{2}]$ rather than the usual interval of periodicity $(0, P]$. Since both the signal $x_P(t)$ and complex exponentials

$$\exp(-j2\pi m f_0 s) = \cos(2\pi m f_0 s) + j\sin(2\pi m f_0 s)$$

are periodic with period P, any interval of length P will do.

Instead of considering aperiodic signals right off the bat, we will make a gradual transition from the analysis of periodic to aperiodic signals by considering what happens with the Fourier series if in the above representation (2.3.1) period P increases to ∞; the limit case of infinite period $P = \infty$ would then correspond to the case of an aperiodic signal.

To see the limit behavior of the Fourier series (2.3.1), we shall introduce the following notation:

(1) The multiplicities of the fundamental frequency will become a running discrete variable f_m:

$$f_m = m \cdot f_0;$$

(2) The increments of the new running variable will be denoted by

$$\Delta f_m = f_m - f_{m-1} = f_0 = \frac{1}{P}.$$

In this notation the Fourier expansion (2.2.1) can be rewritten in the form

$$x_P(t) = \sum_{m=-\infty}^{\infty} \left[\int_{-P/2}^{P/2} x(s)e^{-j2\pi f_m s}\,ds \right] e^{j2\pi f_m t}\,\Delta f_m \tag{2.3.2}$$

because $\Delta f_m = f_0 = \frac{1}{P}$. Now, if the period $P \to \infty$, which is the same as assuming that the fundamental frequency $f_0 = \Delta f_m \to 0$, the sum on the right-hand side of the formula (2.3.2) converges to the integral so that our Fourier representation (2.3.2) of a periodic signal $x_P(t)$ becomes the following integral identity for the aperiodic signal:

$$x_\infty(t) = \int_{-\infty}^{\infty} \left[\int_{-\infty}^{\infty} x_\infty(s)e^{-j2\pi f s}\,ds \right] e^{j2\pi f t}\,df. \tag{2.3.3}$$

The inner transformation

$$X(f) = \int_{-\infty}^{\infty} x(t)e^{-j2\pi f t}\,dt \tag{2.3.4}$$

---

is called the *Fourier transform* of signal $x(t)$, and the outer transform

$$x(t) = \int_{-\infty}^{\infty} X(f)e^{j2\pi ft}df \tag{2.3.5}$$

is called the *inverse Fourier transform* of (complex in general) function $X(f)$. The variable in the Fourier transform is the frequency f.

Note that since $|e^{-j2\pi ft}| = 1$, the necessary condition for the existence of the Fourier transform in the usual sense is the absolute integrability of the signal:

$$\int_{-\infty}^{\infty} |x(t)|dt < \infty. \tag{2.3.6}$$

Later on we will try to extend its definition to some important nonintegrable signals.

Example 2.3.1. Let us trace the above limit procedure in the case of an aperiodic signal $x_\infty(t) = e^{-|t|}$. If this signal is approximated by periodic signals with period P obtained by truncating $x(t)$ to the interval $[-\frac{P}{2}, \frac{P}{2})$ and extending it periodically, i.e.,

$$x_P(t) = e^{-|t|} \quad \text{for } -\frac{P}{2} \le t < \frac{P}{2},$$

then the Fourier coefficients of the latter are, remembering that $P = \frac{1}{f_0}$,

$$\begin{aligned} z_{m,P} &= \frac{1}{P}\int_{-P/2}^{P/2} e^{-|t|}e^{-j2\pi mt/P}dt \\ &= \frac{2f_0}{1+(2\pi mf_0)^2}(1 - e^{-1/(2f_0)}(\cos(2\pi mf_0) \\ &\qquad + 2\pi mf_0 \sin(2\pi mf_0))). \end{aligned}$$

Since the original periodic signal $x_P(t)$ was even, the Fourier coefficients $z_m = z_{-m}$, so that the discrete spectrum of $x_P(t)$ is symmetric. Now, as $P \to \infty$, that is $f_0 = \frac{1}{P} \to 0$, the exponentional term $e^{-1/(2f_0)} \to 0$, and with $f_0 = \Delta f$, $mf_0 = f$, we get that

$$z_{m,P} \to \frac{2}{1+(2\pi f)^2}df.$$

Thus the Fourier transform of the aperiodic signal $x_\infty(t)$ is

$$X_\infty(f) = \frac{2}{1+(2\pi f)^2}.$$

Taking the inverse Fourier transform, we verify[5] that

$$\int_{-\infty}^{\infty} \frac{2}{1+(2\pi f)^2}e^{j2\pi ft}df = e^{-|t|}.$$

[5] When faced with integrals of this sort, the reader is advised to consult a book of integrals, or a computer package such as *Mathematica* or MAPLE.

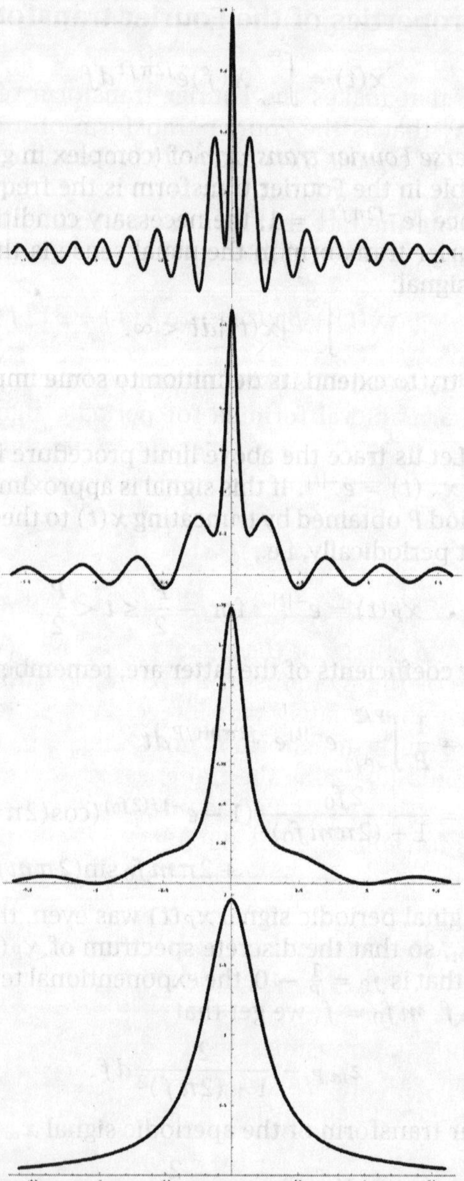

Fig. 2.3.1. Adjusted Fourier coefficients $Z_{m,P} \cdot P$, shown as functions of continuous parameter m for graphical convenience, approach the Fourier transform $X_\infty(f)$ of the aperiodic signal $x_\infty(t) = e^{-|t|}$. The values of P, from top to bottom, are 1, 2, 4, 8.

Figure 2.3.1 illustrates the convergence, as period P increases, of Fourier coefficients $z_{m,P}$ to the Fourier transform $X_\infty(f)$.

2.4 Basic properties of the Fourier transform

The property that makes the Fourier transform of signals so useful is its *linearity*, that is the Fourier transform of a linear composition $\alpha x(t) + \beta y(t)$ of signals $x(t)$ and $y(t)$ is the same linear composition $\alpha X(f) + \beta Y(f)$ of their Fourier transforms. To facilitate notation we will often denote the fact that $X(f)$ is the Fourier transform of signal $x(t)$ by writing $x(t) \mapsto X(f)$. So

$$\alpha x(t) + \beta y(t) \longmapsto \alpha X(f) + \beta Y(f). \tag{2.4.1}$$

The proof is instantaneous using linearity of the integral.

The familiar Parseval formula for periodic signals carries over in the form

$$E_x = \int_{-\infty}^{\infty} |x(t)|^2 dt = \int_{-\infty}^{\infty} |X(f)|^2 df. \tag{2.4.2}$$

That is, the total energy of the signal can be calculated as the integral of the square of the modulus of its Fourier transform. An observant reader will see immediately that integrability of the signal necessary to define the Fourier transform is not sufficient for the validity of the Parseval formula (2.4.2) as the finiteness of the integral $\int_{-\infty}^{\infty} |x(t)| dt$ does not imply that the signal has finite energy E_x.

Parseval's formula also has the following useful extension:

$$\int_{-\infty}^{\infty} x(t) \cdot y(t) dt = \int_{-\infty}^{\infty} X(f) \cdot Y^*(f) df. \tag{2.4.3}$$

In the context of transmission of signals through linear systems the critical property of the Fourier transform is that the *convolution* $[x * y](t)$ of signals $x(t)$ and $y(t)$,

$$[x * y](t) = \int_{-\infty}^{\infty} x(s) y(t-s) ds, \tag{2.4.4}$$

a fairly complex operation, has the Fourier transform that is simply the product of the corresponding Fourier transforms

$$[x * y](t) \longmapsto X(f) \cdot Y(f). \tag{2.4.5}$$

Indeed,

Table 2.4.1. Fourier transform properties.

Signal		Fourier Transform
	Linearity	
$\alpha x(t) + \beta y(t)$	$\longmapsto$	$\alpha X(f) + \beta Y(f)$
	Convolution	
$[x * y](t)$	$\longmapsto$	$X(f) \cdot Y(f)$
	Differentiation	
$x^{(n)}(t)$	$\longmapsto$	$(j2\pi f)^n X(f)$
	Time reversal	
$x(-t)$	$\longmapsto$	$X(-f)$
	Time delay	
$x(t-t_0)$	$\longmapsto$	$X(f) \cdot e^{-j2\pi t_0 f}$
	Frequency translation	
$x(t) \cdot e^{j2\pi f_0 t}$	$\longmapsto$	$X(f-f_0)$
	Frequency differentiation	
$(-j)^n t^n x(t)$	$\longmapsto$	$(2\pi)^{-1} X^{(n)}(f)$
	Frequency convolution	
$x(t)y(t)$	$\longmapsto$	$[X * Y](f)$

$$
\begin{aligned}
\int_{-\infty}^{\infty} &[x * y](t) e^{-j2\pi f t} dt \\
&= \int_{-\infty}^{\infty} \left[\int_{-\infty}^{\infty} x(s) y(t-s) ds \right] e^{-j2\pi f t} dt \\
&= \int_{-\infty}^{\infty} \int_{-\infty}^{\infty} y(t-s) e^{-j2\pi f (t-s)} x(s) e^{-j2\pi f s} ds dt \\
&= \int_{-\infty}^{\infty} y(u) e^{-j2\pi f u} du \cdot \int_{-\infty}^{\infty} x(s) e^{-j2\pi f s} ds = X(f) \cdot Y(f),
\end{aligned}
$$

where the penultimate equality resulted from the substitution $t - s = u$.

Since many electrical circuits are described by differential equations, the behavior of the Fourier transform under differentiation of the signal is another important issue. Here the calculation is also direct:

$$\int_{-\infty}^{\infty} x'(t)e^{-j2\pi ft}dt = x(t)e^{-j2\pi ft}|_{-\infty}^{\infty} + j2\pi f\int_{-\infty}^{\infty} x(t)e^{-j2\pi ft}dt$$
$$= 0 + j2\pi f X_Z(f).$$

The first term is 0 because the signal's absolute integrability (remember, we have to assume it to guarantee the existence of the Fourier transform) implies that $x(\infty) = x(-\infty) = 0$. Thus we have a rule

$$x'(t) \longrightarrow (j2\pi f)\cdot X(f). \qquad (2.4.6)$$

The above and other, simple-to-derive rules are summarized in Table 2.4.1.

Example 2.4.1. Consider the signal $x(t) = e^{-\pi t^2}$, which has the familiar bell shape. Its Fourier transform is

$$X(f) = \int_{-\infty}^{\infty} e^{-\pi t^2 - j2\pi ft}dt = \int_{-\infty}^{\infty} e^{-\pi(t+jf)^2}e^{-\pi f^2}dt = e^{-\pi f^2},$$

because $\int_{-\infty}^{\infty} e^{-\pi(t+jf)^2}dt = \int_{-\infty}^{\infty} e^{-\pi t^2}dt = 1$. Indeed, changing to polar coordinates r, θ, we can evaluate easily that

$$\left(\int_{-\infty}^{\infty} e^{-\pi t^2}dt\right)^2 = \int_{-\infty}^{\infty} e^{-\pi t^2}dt \cdot \int_{-\infty}^{\infty} e^{-\pi s^2}ds$$
$$= \int_{-\infty}^{\infty}\int_{-\infty}^{\infty} e^{-\pi(t^2+s^2)}dtds = \int_0^{2\pi} d\theta \int_0^{\infty} e^{-\pi r^2}rdr = 1.$$

Thus the signal $x(t) = e^{-\pi t^2}$ has the remarkable property of having the Fourier transform of exactly the same functional shape. This fact has profound consequences in mathematical physics and Fourier analysis.

2.5 Fourier transforms of some nonintegrable signals; Dirac delta impulse

There exist important nonintegrable signals, such as $x(t)$ = constant, or $x(t) = \cos t$ that are not absolutely integrable over the whole time-line, and the usual calculus does not permit us to define their Fourier transforms. However, to cover these and other important cases, one can extend the standard calculus by introduction of the so-called *Dirac delta "function"* $\delta(f)$ which is an infinitely high but infinitely narrow spike located at $f = 0$ which, very importantly, has the "area," that is the "integral," equal to 1.[6]

Intuitively, but one can also make this approach rigorous, the best way to think about the Dirac delta is as a limit

$$\delta(f) = \lim_{\epsilon \to 0} r_\epsilon(f), \tag{2.5.1}$$

where

$$r_\epsilon(f) = \begin{cases} \frac{1}{2\epsilon} & \text{for } -\epsilon \le f \le +\epsilon; \\ 0 & \text{elsewhere} \end{cases}$$

is a family, indexed by ϵ, of rectangular functions all of which have area 1 underneath; see Figure 2.5.1.

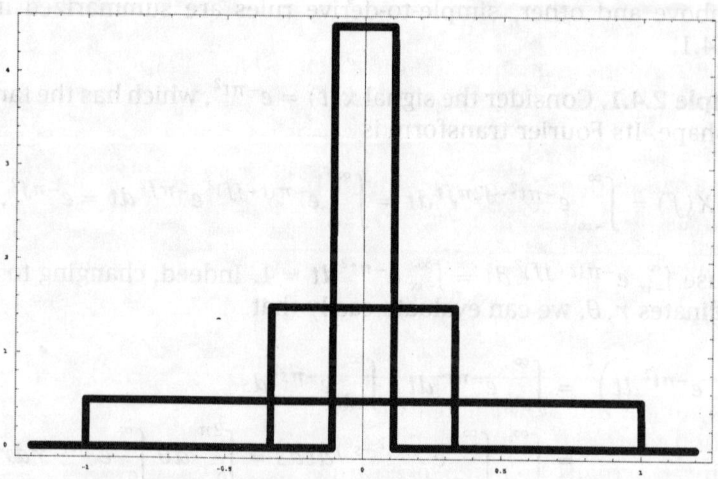

Fig. 2.5.1. Approximation of the Dirac delta $\delta(f)$ by rectangular functions $r_\epsilon(f)$ for $\epsilon = 1, \frac{1}{3}$, and $\frac{1}{9}$.

Obviously the choice of the rectangular functions is not unique here. Any sequence of nonnegative functions which integrate to 1 over the whole real line and converge to zero pointwise at every point different from the origin would do. For example, as approximations to the Dirac delta we can also take the family of double-sided exponential functions of variable x,

$$\frac{1}{2a} \exp\left(\frac{|f|}{a}\right),$$

indexed by parameter $a \to 0+$. Three functions of this family, for parameter values $a = 1, \frac{1}{3}, \frac{1}{9}$, are pictured in Figure 2.5.2.

---

[6] Of course, one can similarly introduce the time domain Dirac delta $\delta(t)$, in which case it will be called the *Dirac delta impulse*.

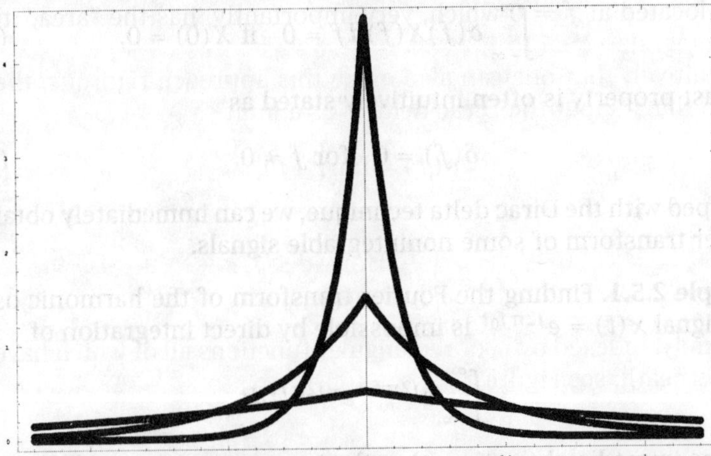

Fig. 2.5.2. Approximation of the Dirac delta $\delta(f)$ by two-sided exponential functions $(\frac{1}{2a})\exp(-\frac{|f|}{a})$ for $a = 1, \frac{1}{3}$, and $\frac{1}{9}$.

The Dirac delta is characterized by its "probing property" (also known as the "sifting property"):

$$\int_{-\infty}^{\infty} \delta(f)X(f)\,df = X(0); \tag{2.5.2}$$

integrating a function against the Dirac delta produces a value of the function at $f = 0$. Operationally, all we need is the formula (2.5.2), which can actually be taken as a formal definition of the Dirac delta.

The "probing" formula (2.5.2) can be justified by remembering our intuitive definition (2.5.1): Indeed, if function $X(f)$ is regular enough, then

$$\begin{aligned}\int_{-\infty}^{\infty} \delta(f)X(f)\,df &= \lim_{\epsilon\to 0}\int_{-\infty}^{\infty} r_\epsilon(f)X(f)\,df \\ &= \lim_{\epsilon\to 0}\frac{1}{2\epsilon}\int_{-\epsilon}^{\epsilon} X(f)\,df = X(0)\end{aligned}$$

in view of the fundamental theorem of calculus.

Other properties of the Dirac delta follow immediately:

$$\int_{-\infty}^{\infty} \delta(f-f_0)X(f)\,df = X(f_0), \tag{2.5.3}$$

$$\int_{-\epsilon}^{\epsilon} \delta(f)\,df = 1, \tag{2.5.4}$$

and

$$\int_{-\infty}^{\infty} \delta(f)X(f)df = 0 \quad \text{if } X(0) = 0, \tag{2.5.5}$$

The last property is often intuitively stated as

$$\delta(f) = 0 \quad \text{for } f \neq 0. \tag{2.5.6}$$

Equipped with the Dirac delta technique, we can immediately obtain the Fourier transform of some nonintegrable signals.

Example 2.5.1. Finding the Fourier transform of the harmonic oscillation signal $x(t) = e^{j2\pi f_0 t}$ is impossible by direct integration of

$$\int_{-\infty}^{\infty} e^{j2\pi f_0 t} e^{-j2\pi f t} dt.$$

But one immediately notices that the inverse transform of the shifted Dirac delta is, by (2.5.2),

$$\int_{-\infty}^{\infty} \delta(f - f_0) e^{j2\pi f t} df = e^{j2\pi f_0 t}.$$

Thus the Fourier transform of $x(t) = e^{j2\pi f_0 t}$ is $\delta(f - f_0)$. In particular, the Fourier transform of a constant 1 is $\delta(f)$ itself.

Table 2.5.1 lists Fourier transforms of some common signals. Here and subsequently, $u(t)$ denotes Heaviside's unit step function, equal to 0 for $t < 0$ and 1 for $t \geq 0$.

Example 2.5.2. The Fourier transform of the signal $x(t) = \cos 2\pi t$ can be found in a similar fashion, as direct integration of

$$\int_{-\infty}^{\infty} \cos(2\pi t) e^{-j2\pi f t} dt$$

is impossible. But one observes that the inverse transform

$$\int_{-\infty}^{\infty} \frac{1}{2}(\delta(f-1) + \delta(f+1)) e^{j2\pi f t} df = \frac{e^{j2\pi t} + e^{-j2\pi t}}{2} = \cos 2\pi t,$$

so the Fourier transform of $\cos 2\pi t$ is $\frac{\delta(f-1)+\delta(f+1)}{2}$.

A sample of the calculus of Dirac delta "functions." There exists a large theory of Dirac delta "functions," and of similar mathematical objects called distributions (in the sense of Schwartz),[7] which develops

[7] For a more complete exposition of the theory and applications of the Dirac delta and related "distributions," see A. I. Saichev and W. A. Woyczyński, *Distributions in the Physical and Engineering Sciences, Vol. 1: Distributional Calculus, Integral Transforms, and Wavelets*, Birkhäuser Boston, Cambridge, MA, 1998.

Table 2.5.1. Common Fourier transforms.

Signal		Fourier Transform
$e^{-a\|t\|}$	$\longmapsto$	$\dfrac{2a}{a^2+(2\pi f)^2},\quad a>0$
$e^{-\pi t^2}$	$\longmapsto$	$e^{-\pi f^2}$
$\begin{cases}1 & \text{for } \|t\| \le \frac{1}{2};\\ 0 & \text{for } \|t\| > \frac{1}{2}.\end{cases}$	$\longmapsto$	$\dfrac{\sin \pi f}{\pi f}$
$\begin{cases}1-\|t\| & \text{for } \|t\| \le 1;\\ 0 & \text{for } \|t\| > 1.\end{cases}$	$\longmapsto$	$\dfrac{\sin^2 \pi f}{\pi^2 f^2}$
$e^{j2\pi f_0 t}$	$\longmapsto$	$\delta(f-f_0)$
$\delta(t)$	$\longmapsto$	1
$\cos 2\pi f_0 t$	$\longmapsto$	$\dfrac{\delta(f+f_0)+\delta(f-f_0)}{2}$
$\sin 2\pi f_0 t$	$\longmapsto$	$\dfrac{j\delta(f+f_0)-\delta(f-f_0)}{2}$
$u(t)=\begin{cases}0 & \text{for } t<0;\\ 1 & \text{for } t\ge 0.\end{cases}$	$\longmapsto$	$\dfrac{1}{2}\delta(f)+\dfrac{1}{j2\pi f}$
$t\cdot u(t)$	$\longmapsto$	$\dfrac{j}{4\pi}\delta'(f)-\dfrac{1}{4\pi^2 f^2}$
$e^{-at}\cdot u(t)$	$\longmapsto$	$\dfrac{1}{a+j2\pi f},\quad a>0$

tools that help carry out operations such as distributional differentiation, distributional multiplication, etc. To give the reader a little taste of it let us start here with the classical integration-by-parts formula which, for usual, vanishing at $f=\pm\infty$ functions $X(f)$ and $Y(f)$, states that

$$\int_{-\infty}^{\infty} X(f)\cdot Y'(f)df = -\int_{-\infty}^{\infty} X'(f)\cdot Y(f)df. \tag{2.5.7}$$

This identity, applied formally, can be used as the *definition* of the derivative $\delta'(f)$ of the Dirac delta by assigning to it the following probing property:

$$\int_{-\infty}^{\infty} X(f)\cdot \delta'(f)df = -\int_{-\infty}^{\infty} X'(f)\cdot \delta(f)df = -X'(0). \tag{2.5.8}$$

Symbolically, we can write

$$X(f) \cdot \delta'(f) = -X'(f) \cdot \delta(f).$$

In the particular case $X(f) = f$ (here, the function has to be thought of as a limit of functions vanishing at $\pm\infty$), we get

$$f \cdot \delta(f) = -\delta(f),$$

a useful computational formula which can be employed, for example, to justify the next to the last entry in the above table of common Fourier transforms.

2.6 Discrete and fast Fourier transforms

In practice, for many signals we only obtain the value of the signal at discrete times, but we can imagine that the signal continues between these times. Thus we can approximate the integrals involved in calculation of the Fourier transforms in the same way as one does in numerical integration in calculus, using left-handed rectangles, trapezoids, Simpson's rule, etc. We use the simplest approximation, which is equivalent to assuming that the signal is constant between the times at which we sample (and rectangular approximations of the area under the function).

Therefore, suppose that the sampling period is T_s, with the sampling frequency $f_s = \frac{1}{T_s}$, so that the signal's sample is given in the form of a sequence

$$x_k = x(kT_s), \quad k = 0, 1, 2, \ldots, N-1, \tag{2.6.1}$$

and we interpret it as a periodic signal with period

$$P = \frac{1}{f_0} = NT_s = \frac{N}{f_s}, \tag{2.6.2}$$

The integral in formula (2.3.1) approximating the Fourier transform of the signal $x(t)$ at discrete frequencies mf_0, $m = 0, 1, 2, \ldots, N-1$, can now, in turn, be approximated by the sum

$$\begin{aligned} X_m = X(mf_0) &= \frac{1}{P} \sum_{k=0}^{N-1} x(kT_s) e^{-j2\pi m f_0 k T_s} \cdot T_s \\ &= \frac{1}{N} \sum_{k=0}^{N-1} x_k e^{-j2\pi mk/N} \end{aligned} \tag{2.6.3}$$

in view of the relationships (2.6.2). The sequence

$$X_m, \quad m = 0, 1, 2, \ldots, N-1, \tag{2.6.4}$$

is traditionally called the *discrete Fourier transform* (*DFT*) of the signal sample x_k, $k = 0, 1, 2, \ldots, N-1$, described in (2.6.1).

Note that the calculation of the DFT via formula (2.6.3) calls for N^2 multiplications $x_k \cdot e^{-j2\pi mk/N}$, $m, k = 0, 1, 2, \ldots, N-1$. One often says that the formula's *computational* (*algorithmic*) *complexity* is of the order N^2. This computational complexity, however, can be dramatically reduced by cleverly grouping terms in the sum (2.6.3). The technique, which usually is called the *Fast Fourier Transform* (*FFT*), was known to Carl Friedrich Gauss at the beginning of the 19th century, but was rediscovered and popularized by Cooley and Tukey in 1965.[8] We will explain it in the special case when the signal's sample size is a power of 2.

So assume that $N = 2^n$, and let $\omega_N = e^{-j2\pi/N}$. It is called a complex Nth root of unity because $\omega_N^N = 1$. Obviously, for $M = \frac{N}{2}$, we have

$$\omega_{2M}^{(2k)m} = \omega_M^{km}, \qquad \omega_M^{M+m} = \omega_M^m, \quad \text{and} \quad \omega_{2M}^{M+m} = -\omega_{2M}^m. \tag{2.6.5}$$

The crucial observation is to recognize that the sum (2.6.3) can be split into two pieces

$$X_m = \frac{1}{2}(X_m^{\text{even}} + X_m^{\text{odd}} \cdot \omega_{2M}^m), \tag{2.6.6}$$

where

$$X_m^{\text{even}} = \frac{1}{M}\sum_{k=0}^{M-1} x_{2k}\omega_M^{km} \quad \text{and} \quad X_m^{\text{odd}} = \frac{1}{M}\sum_{k=0}^{M-1} x_{2k+1}\omega_M^{km}, \tag{2.6.7}$$

and that, in view of (2.6.5),

$$X_{m+M} = \frac{1}{2}(X_m^{\text{even}} - X_m^{\text{odd}} \cdot \omega_{2M}^m). \tag{2.6.8}$$

As a result, only values X_m, $m = 0, 1, 2, \ldots, M-1 = \frac{N}{2-1}$, have to be calculated by laborious multiplication. The values X_m, $m = M, M+1, \ldots, 2M-1 = N-1$, are simply obtained by formula (2.6.8). The above trick is then repeated at levels $\frac{N}{2^2}, \frac{N}{2^3}, \ldots, 2$. If we denote by $\mathrm{CC}(n)$ the *computational complexity* of the above scheme, that is the number of multiplications required, we see that

$$\mathrm{CC}(n) = 2\,\mathrm{CC}(n-1) + 2^{n-1},$$

with the first term on the right being the result of halving the size of the sample at each step, and the second term resulting from multiplications of X_m^{odd} by ω_{2M}^m in (2.6.6) and (2.6.8). Iterating the above recursive relation, one obtains that

[8] J. W. Cooley and O. W. Tukey, An algorithm for the machine calculation of complex Fourier series, *Math. Comput.*, **19** (1965), 297–301.

$$\mathrm{CC}(n) = 2^{n-1} \log_2 2^n = \frac{1}{2} N \log_2 N, \tag{2.6.9}$$

a major improvement over the N^2 order of the computational complexity of the straightforward calculation of DFT.

2.7 Problems and exercises

2.7.1. Prove that the system of real harmonic oscillations

$$\sin(2\pi m f_0 t), \qquad \cos(2\pi m f_0 t), \quad m = 0, 1, 2, \ldots,$$

form an orthogonal system. Is the system normalized? Use the above information to derive formulas for coefficients a_m, b_m, in the expansion (1.2.4). Model this derivation on (2.1.4).

2.7.2. Using the results from Problem 2.7.1, find formulas for amplitudes c_m and phases θ_m in the expansion (1.2.1).

2.7.3. Find a general formula for the coefficients c_m in the cosine Fourier expansion for the even rectangular waveform $x(t)$ from Example 2.1.1.

2.7.4. Find a general formula for the coefficients b_m in the sine Fourier expansion for the odd rectangular waveform $x(t)$ from Example 2.1.2.

2.7.5. Carry out calculations of Example 2.1.3 in the case of arbitrary period P and amplitude a.

2.7.6. Find three consecutive approximations by finite Fourier sums of the signal $x(t)$ from Example 2.1.3. Graph them and compare the graphs with the graph of the original signal.

2.7.7. Find the complex and real Fourier series for the periodic signal with period P defined by the formula

$$x(t) = \begin{cases} a & \text{for } 0 \le t < \frac{P}{2}; \\ -a & \text{for } \frac{P}{2} \le t < P. \end{cases}$$

In the case $P = \pi$ and $a = 2.5$ produce graphs comparing the signal $x(t)$ and its finite Fourier sums of order 1, 3, and 6.

2.7.8. Find the complex and real Fourier series for the periodic signal with period $P = 1$ defined by the formula

$$x(t) = \begin{cases} 1 - \frac{t}{2} & \text{for } 0 \le t < \frac{1}{2}; \\ 0 & \text{for } \frac{1}{2} \le t < 1. \end{cases}$$

Produce graphs comparing the signal $x(t)$ and its finite Fourier sums of order 1, 3, and 6.

2.7.9. Find the complex and real Fourier series for the periodic signal $x(t) = |\sin t|$. Produce graphs comparing the signal $x(t)$ and its finite Fourier sums of order 1, 3, and 6.

2.7.10. Find the complex and real Fourier series for the periodic signal with period $P = \pi$ defined by the formula

$$x(t) = e^t \quad \text{for } -\frac{\pi}{2} < t \le \frac{\pi}{2}.$$

Produce graphs comparing the signal $x(t)$ and its finite Fourier sums of order 1, 3, and 6.

2.7.11. Find an example of a signal $x(t)$ that is absolutely integrable, i.e., $\int_{-\infty}^{\infty} |x(t)|dt < \infty$ but has infinite energy $E_x = \int_{-\infty}^{\infty} |x(t)|^2 dt$, and conversely, find an example of a signal which has finite energy but is not absolutely integrable.

2.7.12. Provide a detailed verification of Fourier transform properties listed in Table 2.4.1.

2.7.13. Provide a detailed verification of the Fourier transforms table (Table 2.5.1). Utilize the fact that the derivative $\delta'(f)$ of the Dirac delta impulse $\delta(f)$ is defined by the integration-by-parts formula

$$\int_{-\infty}^{\infty} \delta'(f)X(f)df = -\int_{-\infty}^{\infty} \delta(f)X'(f)df$$

for any smooth function $X(f)$.

2.7.14. Find the Fourier transform of the periodic signal

$$x(t) = \sum_{m=-\infty}^{\infty} z_m e^{j2\pi m f_0 t}.$$

2.7.15. Find the Fourier transform of the signal $x(t) = tu(t)$, where $u(t)$ is the unit step function equal to 0 for $t < 0$ and 1 for $t \ge 0$.

2.7.16. Find the Fourier transform of the signals given below. Graph both the signal and its Fourier transform:

(a) $$x(t) = \frac{1}{1+t^2}, \quad -\infty < t < \infty,$$

(b) $$e^{-t^2/2}, \quad -\infty < t < \infty,$$

(c) $$x(t) = \begin{cases} \sin t \cdot e^{-t} & \text{for } t \ge 0; \\ 0 & \text{for } t < 0. \end{cases}$$

(d) $x(t) = y * z(t), \qquad y(t) = u(t) - u(t-1), \qquad z(t) = e^{-|t|}$,

where $u(t)$ is the unit step signal = 0 for negative t and = 1 for $t \ge 0$.

2.7.17. Find the convolution $(x * x)(t)$ if $x(t) = u(t) - u(t-1)$, where $u(t)$ is the unit step function. First, use the original definition of the convolution and then verify your result using the Fourier transform method.

2.7.18. Utilize the Fourier transform (in the space variable z) to find a solution of the diffusion (heat) partial differential equation

$$\frac{\partial u}{\partial t} = \sigma \frac{\partial^2 u}{\partial z^2},$$

for a function $u(t, z)$ satisfying the initial condition $u(0, z) = \delta(z)$. The solution of the above equation is often used to describe the temporal evolution of the density of a diffusing substance.[9]

2.7.19. Assuming the validity of the Parseval formula $\int_{-\infty}^{\infty} |x(t)|^2 dt = \int_{-\infty}^{\infty} |X(f)|^2 df$, prove its extended version $\int_{-\infty}^{\infty} x(t) \cdot y^*(t) dt = \int_{-\infty}^{\infty} X(f) \cdot Y^*(f) df$. *Hint*: In the case of real-valued $x(t)$, $y(t)$, $X(f)$, and $Y(f)$, it suffices to utilize the obvious identity $4xy = (x+y)^2 - (x-y)^2$, but in the general, complex case, first verify, and then apply the following *polarization identity*:

$$4xy = |x + y|^2 - |x - y|^2 + j(|x + jy|^2 - |x - jy|^2).$$

Remember that the modulus square $|z|^2 = zz^*$.

[9] It was the search for solutions to this problem that induced Jean-Baptiste Fourier (born March 21, 1768, in Auxerre, France; died May 16, 1830, in Paris) to introduce in his treatise *Théorie analytique de la chaleur* (*The Analytical Theory of Heat*; 1822), the tools of infinite functional series and integral transforms now known under the names of Fourier series and transforms. Fourier was also known as an Egyptologist and administrator. The modern author of research papers, impatient with delays in publication of his/her work, should find solace in the fact that the appearance of Fourier's great memoir was held up by the referees for 15 years; it was first presented to the Institut de France on December 21, 1807.

3

Random Quantities and Random Vectors

By definition, values of random signals at a given sampling time are random quantities which can be distributed over a certain range of values. The tools for the precise, quantitative description of those distributions are provided by the classical *probability theory*. However natural, its development has to be handled with care since the overly heuristic approach can easily lead to apparent paradoxes.[10] But the basic intuitive idea that for independently repeated experiments, probabilities of their particular outcomes correspond to their relative frequencies of appearance, is correct. Although the concept of probability is more elementary than the concept of cumulative probability distribution functions, we assume that the reader is familiar with the former at the high school level, and start our exposition with the latter, which not only applies universally to all types of data, both discrete and continuous, but also gives us a tool to immediately introduce the probability calculus ideas, including the physically appealing probability density function.

Think here about an electrical engineer whose responsibility is to monitor the voltage on the electrical outlets in the university's circuits laboratory. The record of a month's worth of daily readings on a very sensitive voltmeter may look as follows:

```
109.779, 109.37,  110.733, 109.762, 110.364, 110.73,
109.906, 110.378, 109.132, 111.137, 109.365, 108.968,
111.275, 110.806, 110.99,  111.522, 110.728, 109.689,
111.163, 107.22,  109.661, 108.933, 111.057, 111.055,
112.392, 109.55,  111.042, 110.679, 111.431, 112.06.
```

Surprisingly, the voltage varies from day to day and this variability is visualized in Figure 3.0.1.

In the presence of such uncertainty he may want to get a better idea of how the voltage values are distributed within its range and he is

[10] See, e.g., Problem 3.7.25.

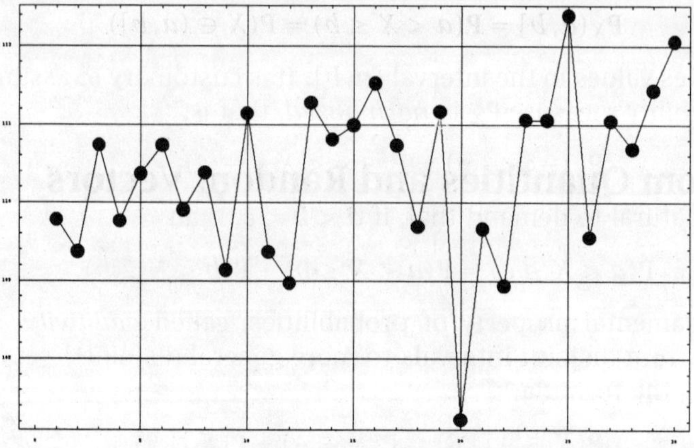

Fig. 3.0.1. Variability of daily voltage readings on an electrical outlet.

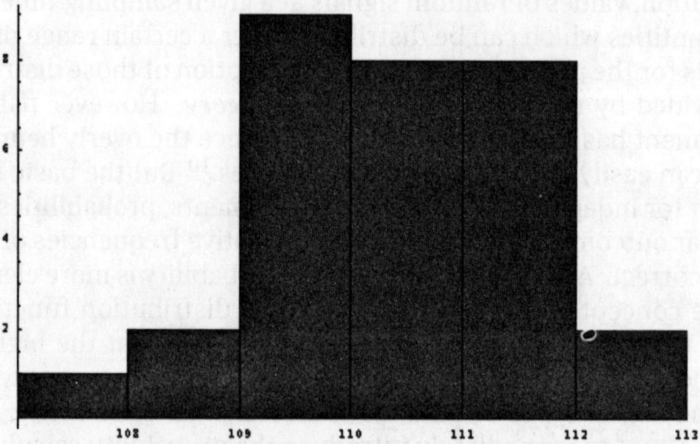

Fig. 3.0.2. The histogram of daily voltage readings on an electrical outlet.

likely to visualize this information in the form of a histogram shown in Figure 3.0.2.

In this chapter, we will discuss analytical tools for the study of such random quantities. The discrete and continuous random quantities are introduced, but we also show that, in the presence of fractal phenomena, the above classification is not exhaustive.

3.1 Discrete, continuous, and singular random quantities

For the purposes of this book, *random quantities* (also called *random variables* in the literature), denoted by capital letters X, Y, etc., will symbolize measurements of experiments with uncertain outcomes. A random quantity X will be fully characterized by its *probability distribution* $\mathbf{P}_X$, which, for any numbers $a < b$, assigns the probability

$$\mathbf{P}_X(a,b] = \mathbf{P}(a < X \le b) = \mathbf{P}(X \in (a,b])$$

that X takes values in the interval $(a,b]$. It is customary to assume that the probability measure $\mathbf{P}_X$ is *normalized*, that is,

$$\mathbf{P}_X(-\infty,+\infty) = \mathbf{P}(-\infty < X < +\infty) = 1, \tag{3.1.1}$$

and it is natural to demand that, if $a < b < c$, than

$$\mathbf{P}(a < X \le c) = \mathbf{P}(a < X \le b) + \mathbf{P}(b < X \le b). \tag{3.1.2}$$

This fundamental property of probabilities, called *additivity*, can be extended from disjoint intervals to more general disjoint[11] sets A and B, yielding the formula

$$\mathbf{P}(X \in A \cup B) = \mathbf{P}(X \in A) + \mathbf{P}(X \in B).$$

In other words, probability measure behaves like the area measure of planar sets.

Equivalently, one can completely characterize the probability distribution $\mathbf{P}$ of X by its *cumulative distribution function* (*c.d.f.*)

$$F_X(x) := \mathbf{P}(X \le x),$$

which gives the probability that the outcomes of experiment X do not exceed number x. Note that, in a sense, c.d.f. $F_X(x)$, which depends only on one variable x, is a simpler object than the probability distribution $\mathbf{P}_X(a,b]$, which depends on two. Properties (3.1.1)–(3.1.2) of $\mathbf{P}$ immediately imply the *normalization* and *monotonicity* of F_X,

$$F_X(-\infty) = 0, \quad x < y \Rightarrow F_X(x) \le F_X(y), \quad F_X(+\infty) = 1, \tag{3.1.3}$$

and the formula recovering $\mathbf{P}$ from F_X:

$$\mathbf{P}(a < X \le b) = F_X(b) - F_X(a). \tag{3.1.4}$$

Discrete probability distributions. A random quantity X with a discrete probability distribution takes on only (finitely or infinitely many) discrete values, say, $x_1, x_2, \ldots$, so that

$$\mathbf{P}(X = x_i) = p_i, \quad i = 1,2,\ldots, \quad 0 < p_i < 1, \quad \sum p_i = 1. \tag{3.1.5}$$

In the discrete case, the c.d.f.

$$F_X(x) = \sum_{i=1}^{\infty} p_i u(x - x_i), \tag{3.1.6}$$

where $u(x)$ is the unit step function. In other words, the c.d.f. has jumps of size p_i at locations x_i, and is constant at other points of the real line.

[11] Recall that sets A and B are called *disjoint* if their intersection is the empty set, i.e., $A \cap B = \emptyset$.

Example 3.1.1 (Bernoulli distribution). In this case the values of X, that is the outcomes of the experiment, are assumed to be either 1 or 0 (think about it as a model of an experiment in which "success" or "failure" are the only possible outcomes), with $\mathbf{P}(X = 1) = p > 0$, $\mathbf{P}(X = 0) = q > 0$, with p, q satisfying condition $p + q = 1$. The c.d.f. of the Bernoulli random quantity is

$$F_X(x) = \begin{cases} 0 & \text{for } x < 0; \\ q = 1 - p & \text{for } 0 \le x < 1; \\ 1 & \text{for } 1 \le x. \end{cases}$$

The Bernoulli family of distributions has one parameter p which must be a number between 0 and 1. Then $q = 1 - p$.

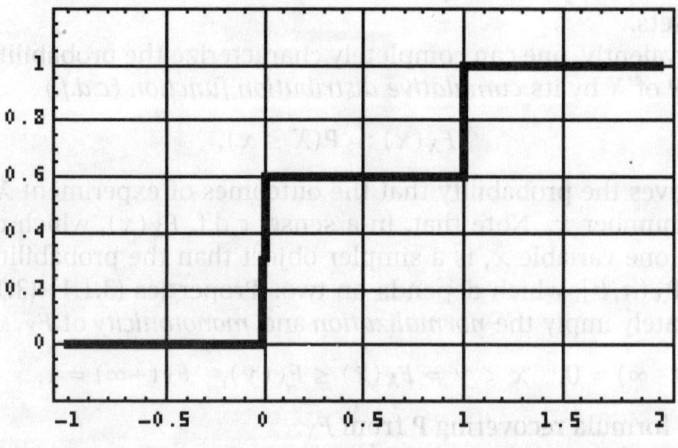

Fig. 3.1.1. Cumulative distribution function $F_X(x)$ of a Bernoulli random quantity X with parameter $p = 0.4$ has a jump of size $q = 1 - 0.4 = 0.6$ at $x = 0$ and a jump of size $p = 0.4$ at $x = 1$.

Example 3.1.2 (binomial distribution). The binomial random quantity X can take values $0, 1, \dots, n$, with corresponding probabilities

$$p_k = \mathbf{P}(X = k) = \binom{n}{k} p^k (1 - p)^{n-k}, \quad k = 0, 1, 2, \dots, n,$$

where the binomial coefficient is defined by

$$\binom{n}{k} = \frac{n!}{k!(n-k)!}.$$

Recall, that the name "binomial coefficient" comes from the elementary *binomial formula*

$$(a+b)^n = \sum_{k=0}^{n} \binom{n}{k} a^k b^{n-k},$$

familiar in the special cases:

$$(a+b)^2 = a^2 + 2ab + b^2,$$
$$(a+b)^3 = a^3 + 3a^2b + 3ab^2 + b^3,$$

and so on.

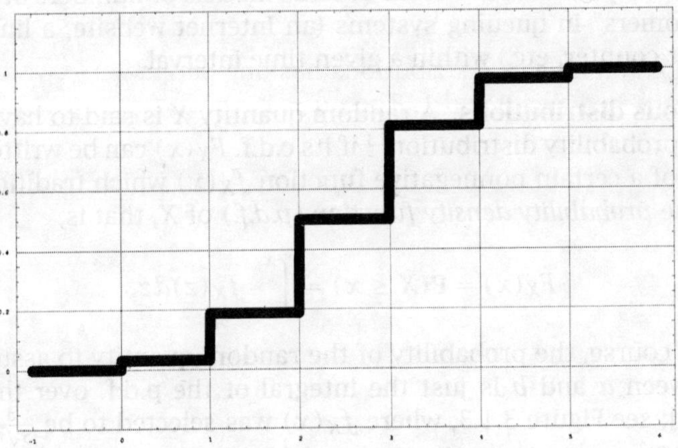

Fig. 3.1.2. Cumulative distribution function $F_X(x)$ of a binomial random quantity X with parameters $p = 0.5$ and $n = 5$.

Probabilities $p_k = p_k(n, p)$ in the binomial probability distribution are probabilities that exactly k "successes" occur in n independent[12] Bernoulli experiments in each of which the probability of "success" is p.

The normalization condition $\sum_k p_k = 1$ (3.1.5) is here satisfied because, in view of the above-mentioned binomial formula,

$$1 = (p+q)^n = \sum_{k=0}^{n} \binom{n}{k} p^k (1-p)^{n-k}.$$

The binomial family of distributions has two parameters: p, which must be between 0 and 1, and n, which can be an arbitrary positive integer.

Example 3.1.3 (Poisson distribution). The values of a Poisson random quantity X can be arbitrary nonnegative integers $0, 1, 2, \ldots$, and their probabilities are defined by the formula

[12] A rigorous definition of the concept of independence of random quantities will be discussed later on in this chapter.

$$p_k = \mathbf{P}(X = k) = e^{-\mu}\frac{\mu^k}{k!}, \quad k = 0, 1, 2, \ldots.$$

The normalization condition $\sum_k p_k = 1$ is satisfied in this case because of the power series expansion for the exponential function:

$$\sum_{k=0}^{\infty} e^{-\mu}\frac{\mu^k}{k!} = e^{-\mu}\sum_{k=0}^{\infty}\frac{\mu^k}{k!} = e^{-\mu}e^{\mu} = 1.$$

The family of Poisson distributions has one parameter $\mu > 0$. Poisson random quantities are often used as models of numbers of arrivals of "customers" in queuing systems (an Internet website, a line at the checkout counter, etc.) within a given time interval.

Continuous distributions. A random quantity X is said to have a continuous probability distribution[13] if its c.d.f. $F_X(x)$ can be written as an integral of a certain nonnegative function $f_X(x)$ which traditionally is called the *probability density function* (*p.d.f.*) of X, that is,

$$F_X(x) = \mathbf{P}(X \le x) = \int_{-\infty}^{x} f_X(z)\,dz. \tag{3.1.7}$$

Then, of course, the probability of the random quantity to assume values between a and b is just the integral of the p.d.f. over the interval $[a, b]$; see Figure 3.1.3, where $f_X(x)$ was selected to be $\frac{3}{5\sqrt{\pi}}e^{-x^2} + \frac{2}{5\sqrt{\pi}}e^{-(x-2)^2}$. Note that in the continuous case it does not matter whether the interval between a and b is open or closed. Thus we have

$$\mathbf{P}(a < X \le b) = F_X(b) - F_X(a) = \int_a^b f_X(z)\,dz. \tag{3.1.8}$$

Also, necessarily, we have the normalization condition

$$\int_{-\infty}^{\infty} f_X(x)\,dx = 1, \tag{3.1.9}$$

and, in view of (3.1.7), and the fundamental theorem of calculus, we can obtain the p.d.f. $f_X(x)$ by differentiation of the c.d.f. $F_X(x)$:

[13] Strictly speaking, c.d.f.s that admit the integral representation (3.1.7), that is, have densities, are called *absolutely continuous distributions* as there exist continuous c.d.f.s which do not admit this integral representation; see an example of a singular c.d.f. later in this section and, e.g., M. Denker and W. A. Woyczyński, *Introductory Statistics and Random Phenomena: Uncertainty, Complexity, and Chaotic Behavior in Engineering and Science*, Birkhäuser Boston, Cambridge, MA, 1998.

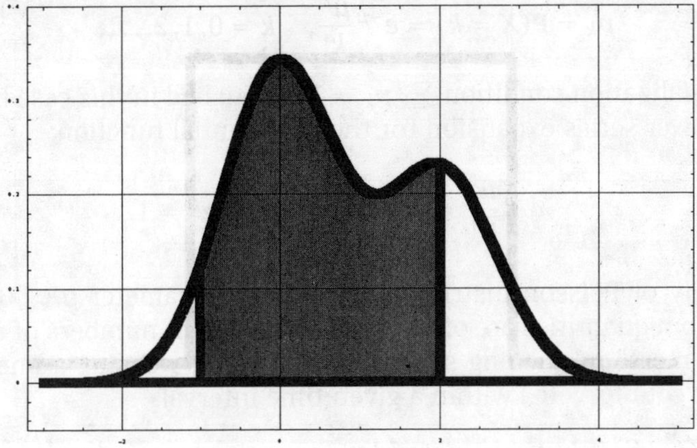

Fig. 3.1.3. The shaded area under $f_X(x)$, and above the interval $[-1, 2]$ is equal to the probability that a random quantity X with p.d.f. $f_X(x)$ takes values in the interval $[-1, 2]$.

$$\frac{d}{dx}F_X(x) = f_X(x).$$

Example 3.1.4 (uniform distribution). The density of a uniformly distributed random quantity X is defined to be a positive constant within a certain interval, say $[c, d]$, and zero outside this interval. Thus, because of the normalization condition (3.1.9),

$$f_X(x) = \begin{cases} (d-c)^{-1} & \text{for } c \le x \le d; \\ 0 & \text{elsewhere.} \end{cases}$$

The family of uniform densities is parametrized by two parameters c and d, with $c < d$.

The c.d.f. of a uniform random quantity is

$$F_X(x) = \begin{cases} 0 & \text{for } x < c; \\ \frac{x-c}{d-c} & \text{for } c \le x \le d; \\ 1 & \text{for } d \le x. \end{cases}$$

Example 3.1.5 (exponential distribution). An exponentially distributed random quantity X has the density of the form

$$f_X(x) = \begin{cases} 0 & \text{for } x < 0; \\ \frac{e^{-x/\mu}}{\mu} & \text{for } x \ge 0. \end{cases}$$

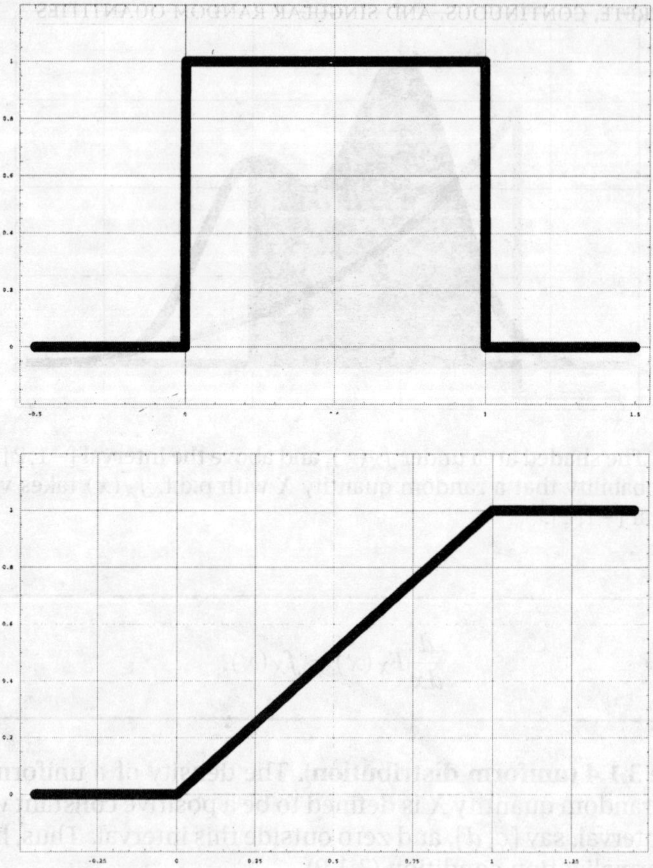

Fig. 3.1.4. *Top*: Probability density function (p.d.f) $f_X(x)$ for a random quantity with values uniformly distributed over the interval $[0, 1]$. *Bottom*: C.d.f. $F_X(x)$ for the same random quantity.

There is one parameter, $\mu > 0$. The c.d.f. in this case is easily computable:

$$F_X(x) = \begin{cases} 0 & \text{for } x < 0; \\ 1 - e^{-x/\mu} & \text{for } x \geq 0. \end{cases}$$

An exponential p.d.f. and the corresponding c.d.f. are pictured in Figure 3.1.3.

Exponential p.d.f.s often appear in applications as probability distributions of random waiting times between Poisson events discussed earlier in this section. For example, under certain simplifying assumptions, it can be proven that the time intervals between consecutive hits of a website have an exponential probability distribution. For this rea-

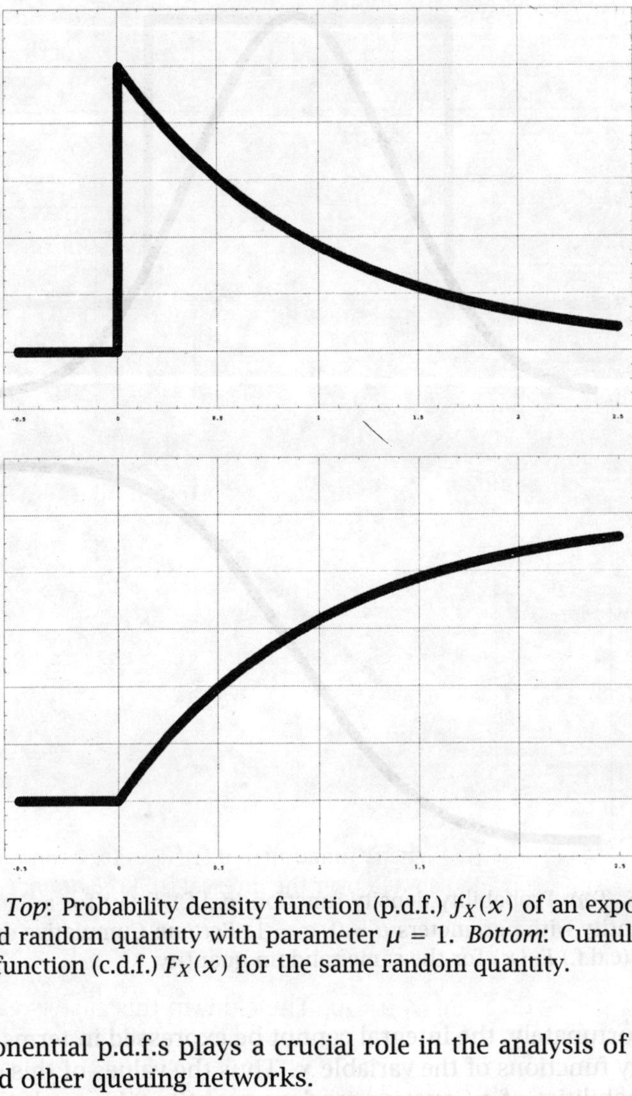

Fig. 3.1.5. *Top*: Probability density function (p.d.f.) $f_X(x)$ of an exponentially distributed random quantity with parameter $\mu = 1$. *Bottom*: Cumulative distribution function (c.d.f.) $F_X(x)$ for the same random quantity.

son, exponential p.d.f.s plays a crucial role in the analysis of Internet traffic and other queuing networks.

Example 3.1.6 (Gaussian (normal) distribution). The density of a Gaussian (also called normal) random quantity X is defined by the formula

$$f_X(x) = \frac{1}{\sqrt{2\pi}\sigma} e^{-(x-\mu)^2/2\sigma^2}.$$

There are two parameters—μ, which is a real number, and $\sigma > 0$—and this distribution is often called the $N(\mu, \sigma^2)$ p.d.f. (N for "normal"). The Gaussian c.d.f. is of the form (see Figure 3.1.4)

$$F_X(x) = \int_{-\infty}^{x} \frac{1}{\sqrt{2\pi}\sigma} e^{-(z-\mu)^2/2\sigma^2} dz,$$

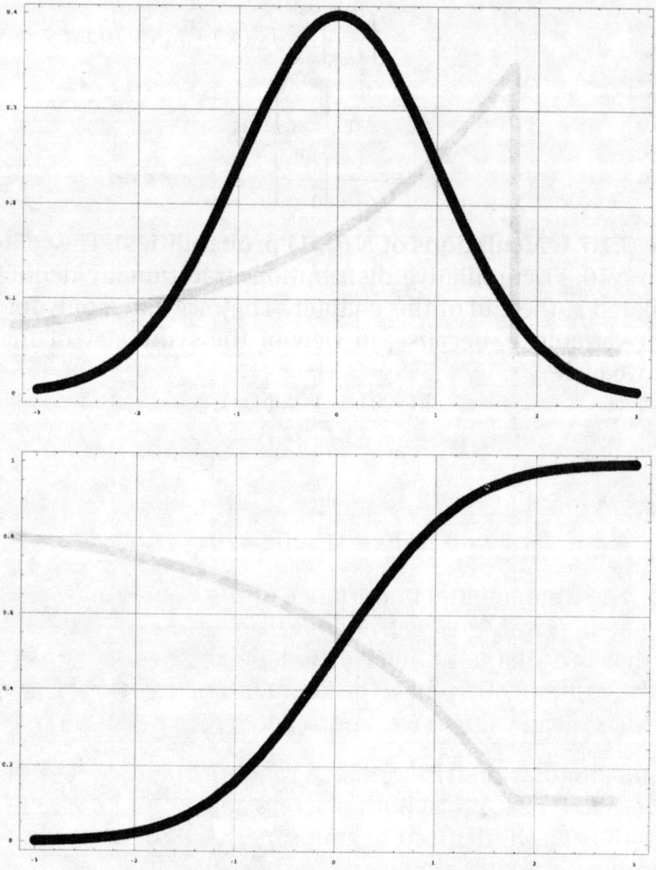

Fig. 3.1.6. *Top*: Probability density function (p.d.f.) $f_X(x)$ for a Gaussian random quantity with parameters $\mu = 0$, $\sigma = 1$. *Bottom*: Cumulative distribution function (c.d.f.) $F_X(x)$ for the same random quantity.

but, unfortunately, the integral cannot be expressed in terms of the elementary functions of the variable x. Thus the values of this c.d.f., and the probabilities of a Gaussian random quantity taking values within a given interval, have to be evaluated numerically, using tables (provided at the end of this chapter), or mathematical software such as MATLAB, MAPLE, or *Mathematica*; see the continuation of Example 3.1.6 below.

However, the normalization condition for the Gaussian p.d.f. can be verified directly analytically by a clever trick that replaces the square of the integral by a double integral which is then evaluated in polar coordinates r, θ. We carry out this calculation in the special case $\mu = 0$, $\sigma^2 = 1$:

$$\left(\int_{-\infty}^{\infty} f_X(x)dx\right)^2 = \int_{-\infty}^{\infty} f_X(x)dx \cdot \int_{-\infty}^{\infty} f_X(y)dy$$

$$= \int_{-\infty}^{\infty}\int_{-\infty}^{\infty} f_X(x)\cdot f_X(y)dxdy$$
$$= \frac{1}{2\pi}\int_{-\infty}^{\infty}\int_{-\infty}^{\infty} e^{-x^2-y^2/2}dxdy$$
$$= \frac{1}{2\pi}\int_{0}^{2\pi}\int_{0}^{\infty} e^{-r^2/2}rdrd\theta = 1.$$

Example 3.1.7 (calculations of $N(0,1)$ probabilities). The values of the Gaussian $N(0,1)$ cumulative distribution, traditionally denoted $\Phi(x)$, are tabulated at the end of this chapter. They are listed only for positive values of variable x, because, in view of the symmetry of the $N(0,1)$ density, we have

$$\Phi(-x) = 1 - \Phi(x).$$

Thus

$$\mathbf{P}(-1.53 < X < 2.11) = \Phi(2.11) - \Phi(-1.53) = \Phi(2.11) - (1 - \Phi(1.53))$$
$$\approx 0.9826 - (1 - 0.9370) = 0.9196.$$

Remark. The fundamental importance of the Gaussian probability distribution stems from the *central limit theorem* (see Section 3.5), which asserts that for a large number of independent repetitions of experiments with random outcomes, the fluctuations (errors) of the outcomes around their mean value have, approximately, a Gaussian p.d.f.

Mixed and singular distributions. A random quantity is said to have a c.d.f. of *mixed type* if it has both discrete and continuous components. The c.d.f. thus has both discrete jumps, perhaps infinitely many, as well as points of continuous increase where its derivative is well defined. For example, the c.d.f.

$$F_X(x) = \begin{cases} 0 & \text{for } x < -1; \\ \frac{x}{6} + \frac{2}{6} & \text{for } -1 \le x < 0; \\ \frac{x}{6} + \frac{4}{6} & \text{for } 0 \le x < 1; \\ 1 & \text{for } 1 \le x, \end{cases} \tag{3.1.10}$$

represents a random quantitiy X which is uniformly distributed on the interval $[-1,1]$ with probability $\frac{1}{3}$, but also takes the discrete values $-1, 0, 1$, with positive probabilites equal to the jump sizes of the c.d.f at those points. Thus, for example,

$$\mathbf{P}\left(-\frac{1}{2} < X \le \frac{1}{2}\right) = F_X\left(\frac{1}{2}\right) - F_X\left(-\frac{1}{2}\right) = \left(\frac{1}{12} + \frac{4}{6}\right) - \left(-\frac{1}{12} + \frac{2}{6}\right) = \frac{1}{2},$$

and

$$\mathbf{P}(X = 0) = \lim_{\epsilon\to 0}\mathbf{P}(-\epsilon < X \le \epsilon) = \lim_{\epsilon\to 0}(F_X(\epsilon) - F_X(-\epsilon))$$

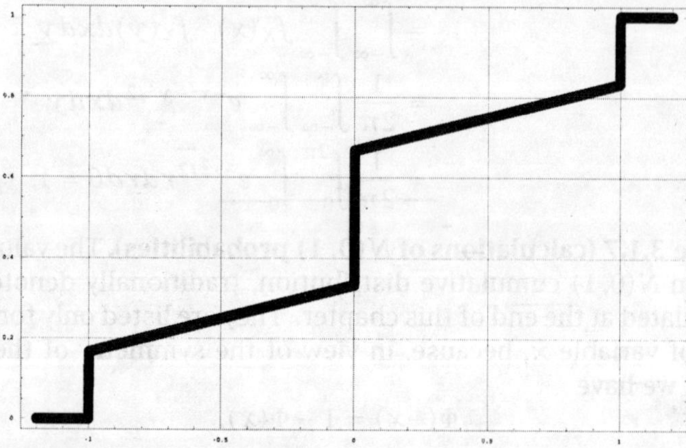

Fig. 3.1.7. Cumulative distribution function (c.d.f.) $F_X(x)$ of mixed type described by formula (3.1.10). This distribution has both discrete and continuous components.

$$= \lim_{\epsilon \to 0} \left[\left(\frac{\epsilon}{6} + \frac{4}{6} \right) - \left(-\frac{\epsilon}{6} + \frac{2}{6} \right) \right] = \frac{1}{3}.$$

Similarly,

$$\mathbf{P}(X = -1) = \frac{1}{6}, \qquad \mathbf{P}(X = 0) = \frac{2}{6}, \qquad \mathbf{P}(X = 1) = \frac{1}{6}.$$

Remark. The reader will notice that the example of a p.d.f. which appeared in Figure 3.1.3 is a mixture of two Gaussian p.d.f.s.

It is tempting to venture a guess that all c.d.f.s have to be either discrete, continuous, or of mixed type. This, however, is not the case.

The limit of the so-called "devil's staircase" c.d.f.s shown in Figure 3.1.8 is an example of a c.d.f. which, although continuous, does not have a p.d.f.

Observe that inside the interval $[0, 1]$ its derivative is 0 on the union of the infinite family of disjoint intervals whose lengths add up to 1. Indeed, as is clear from the construction displayed in Figure 3.1.8, this set has the linear measure

$$\lim_{n \to \infty} \left(\frac{1}{3} + 2 \cdot \frac{1}{3^2} + \cdots + 2^2 \cdot \frac{1}{3^n} \right) = \frac{1}{3} \sum_{i=0}^{\infty} \left(\frac{2}{3} \right)^i = \frac{1}{3} \cdot \frac{1}{1 - \frac{2}{3}} = 1,$$

in view of the formula for the sum of a geometric series. Thus integration of this derivative cannot possibly give a c.d.f. that grows from 0 to 1. Distributions of this type are called *singular* and they arise in studies of fractal phenomena. One can prove that the set of points of

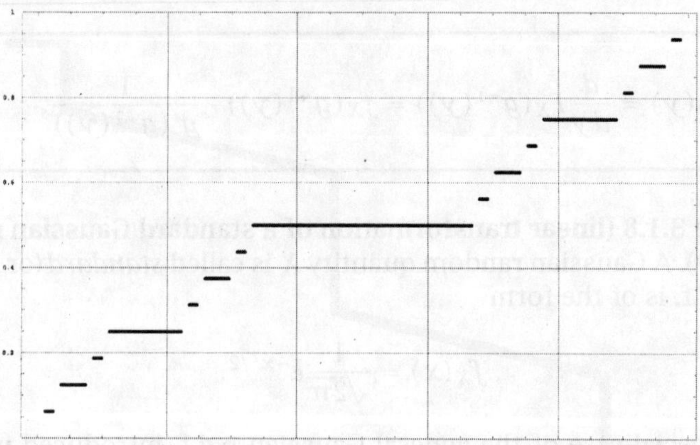

Fig. 3.1.8. The construction of the singular "devil's staircase" c.d.f. $F_X(x)$. It continuously grows from 0 at $x = 0$ to 1 at $x = 1$, and yet it has no density; its derivative is equal to 0 on disjoint intervals whose lengths add up to 1.

increase of the limit "devil's staircase," i.e., the set of points on which the probability is concentrated, has a fractional dimension equal to $\frac{\ln 2}{\ln 3} = 0.6309\ldots$.[14]

Distributions of functions of random quantities. One often measures random quantities through devices that distort the original quantity X to produce a new random quantity, say, $Y = g(X)$, and the natural question is how the c.d.f. $F_X(x)$ of X is affected by such a transformation. In other words, the question is: Can $F_Y(y)$ be expressed in terms of g and $F_X(x)$? In the case when the transforming function $g(x)$ is *monotonically increasing* the answer is simple:

$$F_{g(X)}(y) = \mathbf{P}(g(X) \le y) = \mathbf{P}(X \le g^{-1}(y)) = F_X(g^{-1}(y)), \qquad (3.1.11)$$

where $g^{-1}(y)$ is the inverse function of $g(x)$, that is $g^{-1}(g(x)) = x$, or, equivalently, if $y = g(x)$ then $x = g^{-1}(y)$.

Remembering the chain rule of elementary calculus, and the formula for the derivative of the inverse function $g^{-1}(y)$, we also immediately obtain, in the case of monotonically increasing $g(x)$, the *expression of the p.d.f. of $Y = g(X)$ in terms of the p.d.f. of X itself*:

---

[14] See, for example, M. Denker and W. A. Woyczyński, *Introductory Statistics and Random Phenomena: Uncertainty, Complexity, and Chaotic Behavior in Engineering and Science*, Birkhäuser Boston, Cambridge, MA, 1998.

$$f_{g(X)}(y) = \frac{d}{dy}F_X(g^{-1}(y)) = f_X(g^{-1}(y)) \cdot \frac{1}{g'(g^{-1}(y))}. \qquad (3.1.12)$$

Example 3.1.8 (linear transformation of a standard Gaussian random quantity). A Gaussian random quantity X is called *standard* (or $N(0,1)$) if its p.d.f. is of the form

$$f_X(x) = \frac{1}{\sqrt{2\pi}}e^{-x^2/2}.$$

It is a special case of the general Gaussian p.d.f. introduced in Example 3.1.6, with parameters μ and σ specified to be 0 and 1, respectively. Consider now a new random quantity Y obtained from X by a linear transformation

$$Y = aX + b, \quad a > 0.$$

Think about this transformation as representing the change in units of measurement and the choice of the origin (like changing the temperature measurements from degrees Celsius to Fahrenheit: if X represents temperature measurements in degrees Celsius, then $Y = 1.8 \cdot X + 32$ gives the same measurements in degrees Fahrenheit).

The transforming function in this case, $y = g(x) = ax + b$, is monotonically increasing, and

$$g'(x) = a \quad \text{and} \quad g^{-1}(y) = \frac{y-b}{a}.$$

Formula (3.1.12) now gives the following expression for the p.d.f. of Y:

$$f_Y(y) = 1\sqrt{2\pi}e^{-((y-b)/a)^2/2} \cdot \frac{1}{a} = \frac{1}{\sqrt{2\pi a^2}}e^{-(y-b)^2/2a^2}.$$

The conclusion is that the transformed random quantity Y also has a Gaussian p.d.f., but with parameters $\mu = b$ and $\sigma^2 = a^2$; in other words, Y is $N(b, a^2)$-distributed (in short, $Y \sim N(b, a^2)$).

Example 3.1.9 (calculation of general $N(\mu, \sigma^2)$ probabilities). The relationship established in Example 3.1.8 permits utilization of tables of the $N(0,1)$ distributions supplied at the end of this chapter to calculate $N(\mu, \sigma^2)$ probabilities for arbitrary values of parameter μ and $\sigma > 0$. Indeed, if a random quantity Y has the $N(\mu, \sigma^2)$ distribution, then it is of the form

$$Y = \sigma X + \mu,$$

where X has the $N(0,1)$ distribution, so that

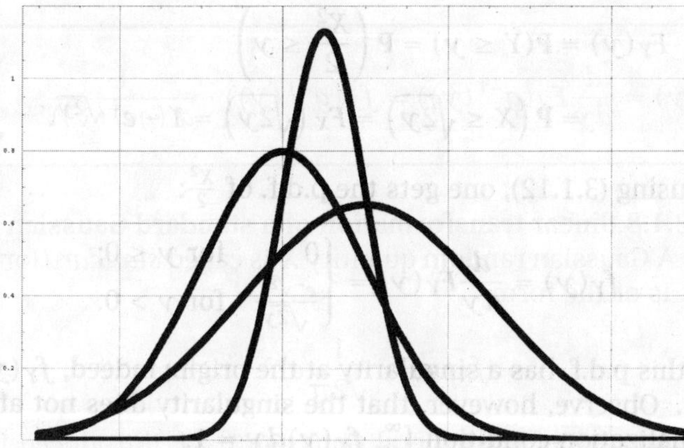

Fig. 3.1.9. Probability density functions of $N(0,1)$, $N(0.5,0.25)$, and $N(1,2.25)$ random quantities (from left to right).

$$
\begin{aligned}
F_Y(y) &= \mathbf{P}(Y \le y) = \mathbf{P}(\sigma X + \mu \le y) \\
&= \mathbf{P}\left(X \le \frac{y-\mu}{\sigma}\right) = \Phi\left(\frac{y-\mu}{\sigma}\right),
\end{aligned} \tag{3.1.13}
$$

and the values of the latter can be taken from the tables. For example, if Y is Gaussian with parameters $\sigma = 1.8$ and $\mu = 32$, then

$$
\begin{aligned}
\mathbf{P}(30 < Y < 36) &= \Phi\left(\frac{36-32}{1.8}\right) - \Phi\left(\frac{30-32}{1.8}\right) \\
&= \Phi(2.22) - (1 - \Phi(-1.11)) \\
&\approx 0.9868 - (1 - 0.8665) = 0.8533.
\end{aligned}
$$

In the next two examples we will consider the quadratic transformation $Y = \frac{X^2}{2}$ corresponding to calculation of the (random) kinetic energy[15] Y of an object of unit mass $m = 1$, traveling with random velocity X.

Example 3.1.10 (kinetic energy of a unit mass traveling with random, exponentially distributed velocity). Suppose that the random quantity X has an exponential c.d.f. and p.d.f. given in Example 3.1.5 with parameter $\mu = 1$. It is transformed by a quadratic "device" $g(x) = \frac{x^2}{2}$ into the random quantity $Y = \frac{X^2}{2}$. Note that the exponential p.d.f. is concentrated on the positive half-line and that the transforming function $g(x)$ is monotonically increasing in that domain. Then the c.d.f. $F_Y(y) = 0$ for $y \le 0$, and for $y > 0$ we can repeat the argument from formula (3.1.11) to obtain

[15] Recall that an object of mass m traveling with velocity v has kinetic energy $E = \frac{mv^2}{2}$.

$$F_Y(y) = \mathbf{P}(Y \le y) = \mathbf{P}\left(\frac{X^2}{2} \le y\right)$$
$$= \mathbf{P}\left(X \le \sqrt{2y}\right) = F_X\left(\sqrt{2y}\right) = 1 - e^{-\sqrt{2y}}.$$

Similarly, using (3.1.12), one gets the p.d.f. of $\frac{X^2}{2}$:

$$f_Y(y) = \frac{d}{dy}F_Y(y) = \begin{cases} 0 & \text{for } y \le 0; \\ \frac{e^{-\sqrt{2y}}}{\sqrt{2y}} & \text{for } y > 0. \end{cases}$$

Note that this p.d.f. has a singularity at the origin; indeed, $f_Y(y) \uparrow +\infty$ as $y \downarrow 0+$. Observe, however, that the singularity does not affect the p.d.f. normalization condition $\int_{-\infty}^{\infty} f_Y(y)dy = 1$.

If the transforming function $y = g(x)$ is *not monotonically increasing* (or decreasing; see Problem 3.7.26 and Sections 8.1–8.2) over the range of the random quantity X (as, for example, $g(x) = x^2$ in the case when X takes both positive and negative values), then a more subtle analysis is required to find the p.d.f. of the random quantity $Y = g(X)$.

Example 3.1.11 (square of a standard Gaussian random quantity). Assume that X has the standard $N(0,1)$ Gaussian p.d.f. and that the transforming function is quadratic: $y = g(x) = x^2$. The quadratic function is monotonically increasing only over the positive half-line; it is monotonically decreasing over the negative half-line. So, we have to proceed with caution, and start with the analysis of the c.d.f. of $Y = X^2$ by taking advantage of the symmetry of the Gaussian p.d.f.:

$$F_Y(y) = \mathbf{P}(Y \le y) = \mathbf{P}(X^2 \le y)$$
$$= 2\mathbf{P}(0 \le X \le \sqrt{y}) = 2\left(F_X(\sqrt{y}) - \frac{1}{2}\right).$$

The above formula, obviously, is valid only for $y > 0$; on the negative half-line the c.d.f. vanishes. Thus the p.d.f. of $Y = X^2$ is

$$f_Y(y) = \frac{d}{dy}F_Y(y) = \begin{cases} 0 & \text{for } y \le 0; \\ \frac{e^{-y/2}}{\sqrt{2\pi y}} & \text{for } y > 0. \end{cases}$$

This p.d.f. is traditionally called the *chi-square* probability density function. We'll see its importance in Section 3.6, where it plays the central role in the statistical parameter estimation problems.

3.2 Expectations and moments of random quantities

The *expected value*, or, in brief, the *expectation* of a random quantity X is its mean value (or, for a physics-minded reader, the center of the probability mass) with different values of X given weights equal to their probabilities. The expectation of X will be denoted $\mathbf{E}X$, or $\mathbf{E}(X)$, whichever is more convenient. So for a discrete random quantity X with $\mathbf{P}(X = x_i) = p_i$, $\sum_i p_i = 1$, we have

$$\mathbf{E}X = \sum_i x_i p_i, \tag{3.2.1}$$

and for an (absolutely) continuous random quantity with probability density $f_X(x)$

$$\mathbf{E}X = \int_{-\infty}^{\infty} x f_X(x)\,dx. \tag{3.2.2}$$

More generally, one can consider the expectation of a function $g(X)$ of a random quantity X, which is defined by the formulas

$$\mathbf{E}[g(X)] = \begin{cases} \sum_i g(x_i) p_i & \text{in the discrete case;} \\ \int_{-\infty}^{\infty} g(x) f_X(x)\,dx & \text{in the continuous case.} \end{cases} \tag{3.2.3}$$

In particular, if $g(x) = x^k$, $k = 1, 2, \ldots$, then the numbers

$$\mu_k(X) = \mathbf{E}g(X) = \mathbf{E}X^k = \begin{cases} \sum_i x_i^k p_i & \text{in the discrete case;} \\ \int_{-\infty}^{\infty} x^k f_X(x)\,dx & \text{in the continuous case} \end{cases} \tag{3.2.4}$$

are called *kth moments* of X. The first moment $\mu_1 = \mu_1(X)$ is just the expectation of $\mathbf{E}X$ of the random quantity X.

If $g(x) = |x|^\alpha$, $-\infty < \alpha < \infty$, then

$$m_k(X) = \mathbf{E}|X|^\alpha$$

are called *αth absolute moments*, and for $g(x) = |x - \mu_1|^\alpha$, the numbers

$$\mathbf{E}|X - \mu_1|^\alpha = \mathbf{E}|X - \mathbf{E}X|^\alpha$$

are called αth central moments of X. The latter measure the mean value of the αth power of the deviation of the random quantity X from its expectation $\mathbf{E}X$. In other words, they provide a family of parameters which measure how the values of the random quantity are spread around its "center of mass." In the special case $\alpha = 2$, the second central moment

$$\mathbf{E}(X - \mathbf{E}X)^2 = \begin{cases} \sum_i (x_i - \mu_1)^2 p_i & \text{in discrete case;} \\ \int_{-\infty}^{\infty} (x - \mu_1)^2 f_X(x)\,dx & \text{in continuous case} \end{cases} \tag{3.2.5}$$

is called the *variance* of the random quantity X and denoted $\mathrm{Var}(X)$. Again, for a physics-minded reader, it is worth noticing that the variance is just the moment of inertia of the probability mass distribution. A simple calculation gives the formula

$$\mathrm{Var}(X) = \mathrm{E}X^2 - (\mathrm{E}X)^2, \tag{3.2.6}$$

which is sometimes simpler computationally than (3.2.5); the variance is thus the difference between the second moment (sometimes also called the *mean square* of a random quantity) and the square of the first moment. This rule is then often phrased as follows: *Variance is equal to the mean square minus the squared mean.*

Example 3.2.1 (moments of the Bernoulli distribution). For the Bernoulli random quantity X, with distribution given in Example 3.1.1, all the moments are

$$\mu_k(X) = 1^k \cdot p + 0^k \cdot (1-p) = p,$$

and the variance is

$$\mathrm{Var}(X) = (1-p)^2 p + (0-p)^2(1-p) = p(1-p).$$

Example 3.2.2 (mean and variance of the uniform distribution). A uniformly distributed random quantity X (see Example 3.1.4) has expectation

$$\mathrm{E}X = \int_c^d x \frac{1}{d-c} dx = \frac{d+c}{2}.$$

Its variance is

$$\mathrm{Var}(X) = \int_c^d \left(x - \frac{d+c}{2}\right)^2 \frac{1}{d-c} dx = \frac{(d-c)^2}{12}.$$

Notice that the *expectation* or *expected value* EX *of a random quantity* X *scales linearly*, that is,

$$\mathrm{E}(\alpha X) = \alpha \mathrm{E}(X), \quad -\infty < \alpha < \infty, \tag{3.2.7}$$

so that the change of scale of the measurements affects the expectations proportionally: if, for example, X is measured in meters, then EX is also measured in meters. Indeed, in the continuous case,

$$\mathrm{E}(\alpha X) = \int_{-\infty}^{\infty} (\alpha x) f_X(x) dx = \alpha \int_{-\infty}^{\infty} x f_X(x) dx = \alpha \mathrm{E}(X),$$

and the discrete case can be verified in an analogous fashion.

On the other hand, the *variance* $\mathrm{Var}(X)$ *has a quadratic scaling*

$$\mathrm{Var}(\alpha X) = \alpha^2 \,\mathrm{Var}(X). \tag{3.2.8}$$

This follows immediately from the linearity of the expectations (3.2.7) and formula (3.2.6). Thus the mean-square deviation has a somewhat unpleasant nonlinear property which implies that if X is measured, say, in meters, then its variance is measured in meters squared.

For this reason, one often considers the *standard deviation* $\mathrm{Std}(X)$ of a random quantity X which is defined as the square root of the variance:

$$\mathrm{Std}(X) = \sqrt{\mathrm{Var}(X)}. \tag{3.2.9}$$

The standard deviation scales linearly, at least for positive α, since

$$\mathrm{Std}(\alpha X) = |\alpha| \,\mathrm{Std}(X), \quad -\infty < \alpha < \infty. \tag{3.2.10}$$

This means that changing the measurement units affects the standard deviation proportionately as well. If a random quantity is measured in meters, then its standard deviation is also measured in meters.

Additionally, observe that the *expectation is additive with respect to constants*, that is, for any constant β, $-\infty < \beta < \infty$,

$$\mathbf{E}(X + \beta) = \mathbf{E}(X) + \beta. \tag{3.2.11}$$

The verification is again immediate and follows from the additivity property of the integrals (or, in the discrete case, sums):

$$\begin{aligned}\mathbf{E}(X + \beta) &= \int_{-\infty}^{\infty} (x + \beta) f_X(x)\,dx \\ &= \int_{-\infty}^{\infty} x f_X(x)\,dx + \int_{-\infty}^{\infty} \beta f_X(x)\,dx = \mathbf{E}(X) + \beta\end{aligned}$$

because $\int_{-\infty}^{\infty} f_X(x)\,dx = 1$.

Finally, the *variance is invariant under translations*, that is, for any constant β, $-\infty < \beta < \infty$,

$$\mathrm{Var}(X + \beta) = \mathrm{Var}(X). \tag{3.2.12}$$

Indeed,

$$\mathrm{Var}(X+\beta) = \mathbf{E}((X+\beta) - \mathbf{E}(X+\beta))^2 = \mathbf{E}(X+\beta-\mathbf{E}(X)-\beta)^2 = \mathrm{Var}(X).$$

The above properties indicate that any random quantity X can be *standardized* by first centering it and then rescaling it so that the standardized random quantity has expectation 0 and variance 1. Indeed, if

$$Z = \frac{X - \mathbf{E}X}{\mathrm{Std}(X)}, \tag{3.2.13}$$

then it immediately follows from (3.2.10)-(3.2.11) that $\mathbf{E}Z = 0$ and $\sigma^2(Z) = 1$.

Example 3.2.3 (mean and variance of the Gaussian distribution). Let us begin with a random quantity X with the standard $N(0,1)$ p.d.f. Its expectation is

$$\mathbf{E}(X) = \int_{-\infty}^{\infty} x \frac{1}{\sqrt{2\pi}} e^{-x^2/2}\,dx = 0$$

because the integrand is an odd function and is integrated over the interval $(-\infty, \infty)$ which is symmetric about the origin. Its variance is thus just the second moment (mean square) of X, which can be evaluated easily by integration by parts[16]

$$\begin{aligned}\mathrm{Var}(X) &= \int_{-\infty}^{\infty} x^2 \frac{1}{\sqrt{2\pi}} e^{-x^2/2}\,dx = \frac{1}{\sqrt{2\pi}} \int_{-\infty}^{\infty} x \cdot (x e^{-x^2/2})\,dx. \\ &= \frac{1}{\sqrt{2\pi}} \left(-x \cdot e^{-x^2/2} \Big|_{-\infty}^{\infty} + \int_{-\infty}^{\infty} e^{-x^2/2}\,dx \right) = 1,\end{aligned}$$

because $\lim_{x\to\pm\infty} x \cdot e^{-x^2/2} = 0$ and $(\frac{1}{\sqrt{2\pi}})\int_{-\infty}^{\infty} e^{-x^2/2}\,dx = 1$.

Now let us consider a general Gaussian random quantity Y with $N(\mu, \sigma^2)$ p.d.f. In view of Example 3.1.8,

$$Y = \sigma X + \mu.$$

The above properties of the expectation and the variance ((3.2.7)-(3.2.8) and (3.2.11)-(3.2.12)) immediately give

$$\mathbf{E}(Y) = \mathbf{E}(\sigma X + \mu) = \sigma \mathbf{E}(X) + \mu = \mu$$

and

$$\mathrm{Var}(Y) = \mathrm{Var}(\sigma X + \mu) = \mathrm{Var}(\sigma X) = \sigma^2\,\mathrm{Var}(X) = \sigma^2.$$

Thus the parameters μ and σ^2 in the Gaussian $N(\mu, \sigma^2)$ p.d.f. are, simply, its expectation and variance.

---

[16] Recall the integration-by-parts formula: $\int f(x)g'(x)\,dx = f(x)g(x) - \int f'(x)g(x)\,dx$.

3.3 Random vectors, conditional probabilities, statistical independence, and correlations

A *random vector* $\mathbf{X}$ has components $X_1, X_2, \ldots, X_d$, which are scalar random quantities, that is,

$$\mathbf{X} = (X_1, X_2, \ldots, X_d),$$

where d is the dimension of the random vector and its statistical properties are completely determined by its *joint c.d.f.*

$$F_{(X_1,\ldots,X_d)}(x_1, \ldots, x_d) = \mathbf{P}(X_1 \le x_1, \ldots, X_d \le x_d).$$

For the sake of simplicity of notation, we shall consider first the case of dimension $d = 2$, and we shall write $\mathbf{X} = (X, Y)$. In the discrete case, for a random vector $\mathbf{X}$ taking discrete values $\boldsymbol{x} = (x, y)$, the joint probability distribution is

$$\mathbf{P}(\mathbf{X} = \boldsymbol{x}) = \mathbf{P}(X = x, Y = y) = p_{\mathbf{X}}(x, y), \tag{3.3.1}$$

and

$$\sum_{(x,y)} p_{\mathbf{X}}(x, y) = 1. \tag{3.3.2}$$

Example 3.3.1 (a Bernoulli random vector). The random vector (X, Y) takes values $(0,0), (0,1), (1,0), (1,1)$, with the following joint probabilities:

$$p_{(X,Y)}(0,0) = (1-p)^2, \quad p_{(X,Y)}(0,1) = p(1-p),$$
$$p_{(X,Y)}(0,1) = (1-p)p, \quad p_{(X,Y)}(1,1) = p^2.$$

It is easy to check that

$$\sum_{x=0}^{1} \sum_{y=0}^{1} p_{(X,Y)}(x, y) = 1.$$

In the special case $p = \frac{1}{2}$ all four possible values of this random vector are taken with the same probability equal to $\frac{1}{4}$.

A continuous random vector is characterized by its *joint p.d.f.* $f_{(X,Y)}(x, y)$, which is a nonnegative function of two variables x, y, such that

$$\int_{-\infty}^{\infty} \int_{-\infty}^{\infty} f_{(X,Y)}(x, y)\,dx\,dy = 1. \tag{3.3.3}$$

In this case, the probability that the random vector (X, Y) takes values in a certain domain A of the 2D space is calculated by evaluating the double integral of the joint p.d.f. over the domain A:

$$\mathbf{P}((X,Y)\in A)=\iint_A f_{(X,Y)}(x,y)\,dxdy. \tag{3.3.4}$$

For example, if the domain A is a rectangle $[a,b]\times[c,d]=\{(x,y): a\le x\le b, c\le y\le d\}$, then

$$\mathbf{P}((X,Y)\in A)=\mathbf{P}(a\le X\le b, c\le Y\le d)=\int_a^b\int_c^d f_{(X,Y)}(x,y)\,dydx. \tag{3.3.5}$$

If the domain $B=\{(x,y): x^2+y^2\le R^2\}$ is a centered disk of radius R, then

$$\mathbf{P}((X,Y)\in B)=\mathbf{P}(X^2+Y^2\le R^2)=\int_{-R}^{R}\int_{-\sqrt{R^2-x^2}}^{\sqrt{R^2-x^2}} f_{(X,Y)}(x,y)\,dydx. \tag{3.3.6}$$

The graph of a 2D joint p.d.f. is a surface over the (x,y)-plane such that the volume underneath it is equal to 1; see (3.3.3).

Example 3.3.2 (a 2D Gaussian random vector). An example of the 2D Gaussian joint p.d.f. is given by the formula

$$f_{(X,Y)}(x,y)=\frac{1}{2\pi\sigma_x\sigma_y}\exp\left[-\frac{(x-\mu_x)^2}{2\sigma_x^2}-\frac{(y-\mu_y)^2}{2\sigma_y^2}\right]. \tag{3.3.7}$$

where $\sigma_x,\sigma_y>0$, and μ_x, μ_y are arbitrary real numbers. Figure 3.3.1 shows the plot of the surface representing a 2D Gaussian joint p.d.f. in the case $\sigma_x,\sigma_y=1$ and $\mu_x,\mu_y=0$.

Calculation of the probabilities $\mathbf{P}(a\le X\le b, c\le Y\le d)$ is here reduced to calculation of one-dimensional Gaussian probabilities since the joint 2D density in this case is a product of two 1D Gaussian densities, one depending only on x and the other on y,[17] and the double integral splits into a product of two single integrals. To obtain numerical values, tables of (or software for) 1D $N(0,1)$ c.d.f. have to be used; see Section 3.5.

In the special case of equal variances $\sigma_x^2=\sigma_y^2=\sigma^2$, the probability that the above Gaussian random vector takes values in a disk of radius R centered at (μ_x,μ_y) can, however, be carried out explicitly by calculation of the integral in polar coordinates (θ,r):

$$\mathbf{P}((X-\mu_x)^2+(Y-\mu_y)^2\le R^2)$$
$$=\int_{-R}^{R}\int_{-\sqrt{R^2-x^2}}^{\sqrt{R^2-x^2}}\frac{1}{2\pi\sigma^2}\exp\left[-\frac{x^2+y^2}{2\sigma^2}\right]dydx$$

[17] We will have more to say about joint p.d.f.s of this type in the next few pages. The multiplicative property is equivalent to the concept of statistical independence of components of a random vector.

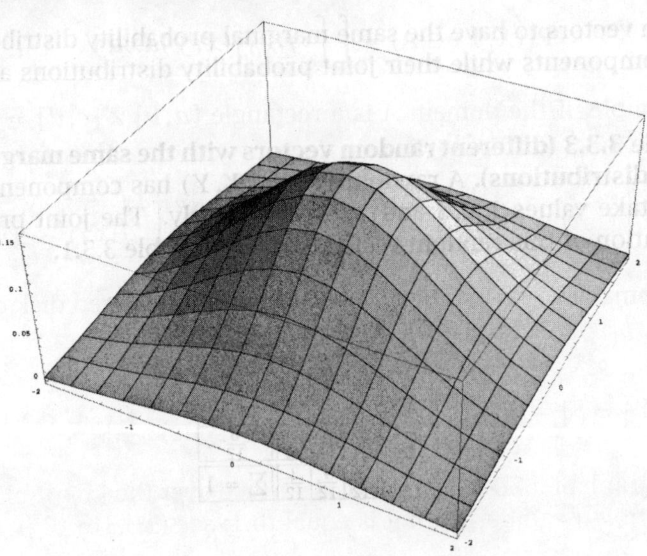

Fig. 3.3.1. Plot of the surface representing a 2D Gaussian joint p.d.f. (3.3.7) in the case $\sigma_x, \sigma_y = 1$ and $\mu_x, \mu_y = 0$.

$$= \frac{1}{2\pi\sigma^2}\int_0^{2\pi}\int_0^R \exp\left[-\frac{r^2}{2\sigma^2}\right] r\,dr\,d\theta$$

$$= \frac{1}{\sigma^2}\left[-\sigma^2\exp\left[-\frac{r^2}{2\sigma^2}\right]\right]_0^R = 1 - e^{-R^2/2\sigma^2}.$$

Because the joint p.d.f. gives complete information about the random vector (X, Y), it also yields complete information about the probability distributions of each of the component random quantities. These distributions are called *marginal distributions* of the random vector. In particular, for a discrete random vector, the marginal distribution of the component X is

$$p_X(x) = \sum_y p_{(X,Y)}(x, y). \tag{3.3.8}$$

To find the probability of X taking a particular value x_0 we simply need to sum, over all possible ys, the probabilities of (X, Y) taking values (x_0, y). For a continuous random vector the marginal p.d.f. of the component X is

$$f_X(x) = \int_{-\infty}^{\infty} f_{(X,Y)}(x, y)\,dy. \tag{3.3.9}$$

It is important to observe that the marginal distributions of components of a random vector do not determine its joint distribution. Indeed, the example provided below shows that it is quite possible for random vectors to have the same marginal probability distributions of their components while their joint probability distributions are different.

Example 3.3.3 (different random vectors with the same marginal probability distributions). A random vector (X, Y) has components X and Y that take values $1, 2, 3$ and $1, 2$, respectively. The joint probability distribution of this random vector is given in Table 3.3.1.

Table 3.3.1.

$Y\backslash X$	1	2	3	Y
1	$\frac{5}{24}$	$\frac{4}{24}$	$\frac{3}{24}$	$\frac{6}{12}$
2	$\frac{5}{24}$	$\frac{4}{24}$	$\frac{3}{24}$	$\frac{6}{12}$
X	$\frac{5}{12}$	$\frac{4}{12}$	$\frac{3}{12}$	$\sum = 1$

Thus, for example, $\mathbf{P}((X, Y) = (3, 2)) = \frac{3}{24}$. The last row in the above table gives the marginal probability distribution for the component X, and the last column, the marginal probability distribution for the component Y.

Now consider another random vector (W, Z) with components W and Z which also take values $1, 2, 3$ and $1, 2$, respectively. The joint distrbution of this random vector is given by Table 3.3.2.

Table 3.3.2.

$Z\backslash W$	1	2	3	Z
1	$\frac{1}{12}$	$\frac{2}{12}$	$\frac{3}{12}$	$\frac{6}{12}$
2	$\frac{4}{12}$	$\frac{2}{12}$	0	$\frac{6}{12}$
W	$\frac{5}{12}$	$\frac{4}{12}$	$\frac{3}{12}$	$\sum = 1$

This time, $\mathbf{P}((X, Y) = (3, 2)) = 0$. The last row in the above table gives the marginal probability distribution for the component W, and the last column, the marginal probability distribution for the component Z. The marginal probability distributions for vectors (X, Y) and (W, Z) are the same, while their joint distributions are different.

Conditional probabilities. Knowledge of the joint p.d.f. permits us also to introduce the concept of the *conditional probability* (in the discrete case) and the *conditional density* (in the continuous case). Thus, the conditional probability of the component X taking value x, given that the second component Y took value y, is given by the formula

$$p_{X|Y}(x|y) \equiv \mathbf{P}(X = x|Y = y) = \frac{\mathbf{P}(X = x, Y = y)}{\mathbf{P}(Y = y)} = \frac{p_{(X,Y)}(x, y)}{p_Y(y)}, \tag{3.3.10}$$

and the conditional probability density function of X given $Y = y$ is given by the formula

$$f_{X|Y}(x|y) = \frac{f_{(X,Y)}(x, y)}{f_Y(y)}. \tag{3.3.11}$$

In other words, conditional probability distributions are distributions of values of one component of a random vector calculated under the assumption that the value of the other component has already been determined.

Conditional probabilities are bona fide probabilities, as they satisfy the normalization property. Indeed, say, in the continuous case, for each fixed y,

$$\int_{-\infty}^{\infty} f_{X|Y}(x|y)dx = \frac{\int_{-\infty}^{\infty} f_{(X,Y)}(x, y)dx}{f_Y(y)} = \frac{f_Y(y)}{f_Y(y)} = 1,$$

in view of formula (3.3.9), which calculates the marginal density from the joint density.

If the component X of random vector (X, Y) takes on distinct values $x_1, x_2, \ldots, x_n$, then the additive property of probabilities immediately gives the following *total probability formula*:

$$\mathbf{P}(Y = y) = \sum_{i=1}^{n} \mathbf{P}(Y = y|X = x_i) \cdot \mathbf{P}(X = x_i).$$

Remark. Heuristically, one can think about conditional probabilities as probabiilities obtained under additional constraints. Think here about the probability of your running into a bear during a hike. If you are hiking in the city park, the probability of the event may be only 0.0001; in Yellowstone the similar conditional probability may be as high as 0.75. Now assume you participate, with 51 of your classmates, in a raffle and the prize is a trip to Yellowstone; the consolation prize is a group hike in the city park. The total probability of your running into a bear would then be $0.0001 \cdot \frac{51}{52} + 0.75 \cdot \frac{1}{52} \approx 0.015$.

One of the corollaries of the total probability formula is the celebrated *Bayes formula for reverse conditional probabilities* which, loosely speaking, computes the conditional probability of X, given Y, in terms of the conditional probabilities of Y, given X:

$$\mathbf{P}(X = x_i | Y = y) = \frac{\mathbf{P}(Y = y | X = x_i) \cdot \mathbf{P}(X = x_i)}{\sum_{i=1}^{n} \mathbf{P}(Y = y | X = x_i) \cdot \mathbf{P}(X = x_i)}.$$

Indeed,

$$\begin{aligned}\mathbf{P}(X = x_i | Y = y) &= \frac{\mathbf{P}(X = x_i, Y = y)}{\mathbf{P}(Y = y)} \cdot \frac{\mathbf{P}(X = x_i)}{\mathbf{P}(X = x_i)}\\ &= \frac{\mathbf{P}(Y = y | X = x_i) \cdot \mathbf{P}(X = x_i)}{\mathbf{P}(Y = y)},\end{aligned}$$

and an application of the total probability formula immediately gives the final result.

Example 3.3.4 (transmission of a binary signal in the presence of random errors). A channel transmits binary symbols 0 and 1 with random errors. The probability that the symbols 0 and 1 appear at the input of the channel are, respectively, 0.45 amd 0.55. Because of transmission errors, if the symbol 0 appears at the input, then the probability of it being received as 0 at the output is 0.95. The analogous conditional probability is 0.9, for the symbol 1 to be received, given that it was transmitted. Our task is to find the reverse conditional probability that the symbol 1 was transmitted given that 1 was received.

The random vector here is (X, Y), where X is the input signal and Y is the output signal. The problem's description contains the following information:

$$\mathbf{P}(X = 0) = 0.45, \qquad \mathbf{P}(X = 1) = 0.55,$$

and

$$\mathbf{P}(Y = 0 | X = 0) = 0.95, \qquad \mathbf{P}(Y = 1 | X = 1) = 0.9,$$

so that

$$\mathbf{P}(Y = 1 | X = 0) = 0.05, \qquad \mathbf{P}(Y = 0 | X = 1) = 0.1.$$

We are seeking $\mathbf{P}(X = 1 | Y = 1)$ and the Bayes formula gives the answer:

$$\begin{aligned}&\mathbf{P}(X = 1 | Y = 1)\\ &= \frac{\mathbf{P}(Y = 1 | X = 1) \cdot \mathbf{P}(X = 1)}{\mathbf{P}(Y = 1 | X = 0) \cdot \mathbf{P}(X = 0) + \mathbf{P}(Y = 1 | X = 1) \cdot \mathbf{P}(X = 1)}\\ &= \frac{0.9 \cdot 0.55}{0.05 \cdot 0.45 + 0.9 \cdot 0.55} \approx 0.9565.\end{aligned}$$

Statistical independence. Components X and Y of a random vector $X = (X, Y)$ are said to be *statistically independent* if the conditional probabilities of X given Y are independent of Y and vice versa. In the discrete case, this means that, for all x and y,

$$\mathbf{P}(X = x | Y = y) = \mathbf{P}(X = x),$$

which is equivalent to the statement that the joint p.d.f. is the product of the marginal p.d.f.s. Indeed, the above independence assumption and the formula defining the conditional probabilities yield

$$\begin{aligned}\mathbf{P}(X = x, Y = y) &= \mathbf{P}_{(X,Y)}(x, y) \\ &= \mathbf{P}_X(x) \cdot \mathbf{P}_Y(y) = \mathbf{P}(X = x) \cdot \mathbf{P}(Y = y).\end{aligned} \tag{3.3.12}$$

In the continuous case, the analogous definition of independence of X and Y can be stated via the multiplicative formula for the joint p.d.f.:

$$f_{(X,Y)}(x, y) = f_X(x) \cdot f_Y(y). \tag{3.3.13}$$

Note that both the 2D Bernoulli distribution of Example 3.3.1 and the 2D Gaussian distribution of Example 3.3.2 have statistically independent components X and Y. Also, components of the random vector (X, Y) in Example 3.3.3 are independent, as the table was actually obtained by multiplying the marginal probabilities in the corresponding rows and columns. However, the components W and Z of random vector (W, Z) in Example 3.3.3 are not statistically independent. To see this, it is sufficient to observe that

$$\mathbf{P}(W = 3, Z = 2) = 0,$$

but

$$\mathbf{P}(W = 3) \cdot \mathbf{P}(Z = 2) = \frac{3}{12} \cdot \frac{6}{12} = \frac{3}{24} \neq 0.$$

Moments of random vectors and correlations. If a random quantity Z is a function of a random vector (X, Y), say,

$$Z = g(X, Y),$$

then as in Section 3.2, we can calculate the expectation of Z using the joint p.d.f. Indeed,

$$\mathbf{E}Z = \sum_x \sum_y g(x, y) p_{(X,Y)}(x, y) \tag{3.3.14}$$

in the discrete case, and

$$\mathbf{E}Z = \int_{-\infty}^{\infty}\int_{-\infty}^{\infty} g(x,y) f_{(X,Y)}(x,y)\,dx\,dy, \tag{3.3.15}$$

in the continuous case.

A mixed second-order moment corresponding to function $g(x,y) = xy$ will play a pivotal role in the analysis of random signals. The number

$$\varphi_{X,Y} = \mathbf{E}(X \cdot Y) \tag{3.3.16}$$

is called the *correlation* of random quantities X and Y. The related parameter corresponding to $g(x,y) = (x - \mu_x)(y - \mu_y)$,

$$\mathrm{Cov}(X,Y) = \mathbf{E}[(X-\mu_X)(Y-\mu_Y)] = \mathbf{E}(XY) - \mathbf{E}(X)\mathbf{E}(Y), \tag{3.3.17}$$

is called the *covariance* of X and Y. Obviously, the covariance of X and X is just the variance of X:

$$\mathrm{Cov}(X,X) = \mathbf{E}[(X-\mu_X)(X-\mu_Y)] = \mathrm{Var}(X). \tag{3.3.18}$$

By the Cauchy-Schwartz inequality,[18]

$$|\mathrm{Cov}(X,Y)| \le \mathrm{Std}(X) \cdot \mathrm{Std}(Y). \tag{3.3.19}$$

This suggests the introduction of yet another parameter for a 2D random vector which is called the *correlation coefficient* of X and Y:

$$\mathrm{Cor}(X,Y) \equiv \rho_{X,Y} = \frac{\mathrm{Cov}(X,Y)}{\mathrm{Std}(X) \cdot \mathrm{Std}(Y)}. \tag{3.3.20}$$

In view of (3.3.19) the correlation coefficient is always contained between -1 and $+1$:

$$-1 \le \rho_{X,Y} \le 1, \tag{3.3.21}$$

and, in view of (3.3.18), if random components X and Y are linearly dependent, that is, $Y = \alpha X$, then the correlation coefficient takes its extreme values

$$\rho_{X,\alpha X} = \pm 1, \tag{3.3.22}$$

[18] Recall that if $\mathbf{a} = (a_1, \dots, a_d)$ and $\mathbf{b} = (b_1, \dots, b_d)$ are two d-dimensional vectors, then the Cauchy-Schwartz inequality says that the absolute value of their scalar (dot) product is not larger than the product of their norms (magnitudes), i.e., $|\langle \mathbf{a}, \mathbf{b}\rangle| \le \|\mathbf{a}\| \cdot \|\mathbf{b}\|$, where $\langle \mathbf{a}, \mathbf{b}\rangle = a_1 b_1 + \cdots + a_d b_d$, and $\|\mathbf{a}\|^2 = a_1^2 + \cdots + a_d^2$; see Section 3.7.

depending on whether α is positive or negative. In those cases we say that the random quantities X and Y are perfectly (positively or negatively) correlated.

The opposite case is that of statistically independent random quantities X and Y. Then, because of the multiplicative property $f_{(X,Y)}(x, y) = f_X(x)f_Y(y)$ (3.3.12)-(3.3.13) of the joint p.d.f., we always have

$$\mathbf{E}(XY) = \int\int xy f_X(x) f_Y(y) dx dy = \mathbf{E}X \cdot \mathbf{E}Y, \tag{3.3.23}$$

so that

$$\mathrm{Cov}(X, Y) = \mathbf{E}(X - \mu_X)(Y - \mu_Y) = \mathbf{E}(X - \mu_X) \cdot \mathbf{E}(Y - \mu_Y) = 0 \tag{3.3.24}$$

and the correlation coefficient $\rho_{X,Y} = 0$. In other words, statistically independent random quantities are always uncorrelated.[19] The correlation coefficient $\rho_{X,Y}$ is often considered as a measure of "independence" of random quantities X and Y; more appropriately, it should be interpreted as a measure of the "linear association" of random quantities X and Y.

Example 3.3.5 (a discrete 2D distribution with nontrivial correlation). Consider the random vector (W, Z) from Example 3.3.3. The expectations of the components are

$$\mathbf{E}W = 1\left(\frac{5}{12}\right) + 2\left(\frac{4}{12}\right) + 3\left(\frac{3}{12}\right) = \frac{11}{6},$$

$$\mathbf{E}Z = 1\left(\frac{6}{12}\right) + 2\left(\frac{6}{12}\right) = \frac{3}{2}.$$

The variances are

$$\mathrm{Var}(W) = \left(1 - \frac{11}{6}\right)^2\left(\frac{5}{12}\right) + \left(2 - \frac{11}{6}\right)^2\left(\frac{4}{12}\right) + \left(3 - \frac{11}{6}\right)^2\left(\frac{3}{12}\right) = \frac{23}{36},$$

$$\mathrm{Var}(Z) = \left(1 - \frac{3}{2}\right)^2\left(\frac{6}{12}\right) + \left(2 - \frac{3}{2}\right)^2\left(\frac{6}{12}\right) = \frac{1}{4}.$$

The expectation of the product is

$$\mathbf{E}(WZ) = (1 \cdot 1)\left(\frac{1}{12}\right) + (2 \cdot 1)\left(\frac{2}{12}\right) + (3 \cdot 1)\left(\frac{3}{12}\right) + (1 \cdot 2)\left(\frac{4}{12}\right) + (2 \cdot 2)\left(\frac{2}{12}\right) + (3 \cdot 2)0 = \frac{5}{2}.$$

Thus the covariance is

$$\mathrm{Cov}(W, Z) = \mathbf{E}(WZ) - \mathbf{E}(W)\mathbf{E}(Z) = \frac{5}{2} - \left(\frac{11}{6}\right)\left(\frac{3}{2}\right) = -\frac{1}{4},$$

[19] The opposite statement is, in general, not true; see Problem 3.7.28.

and, finally, the correlation coefficient of W and Z is

$$\mathrm{Cor}(W,Z) = \frac{\mathrm{Cov}(W,Z)}{\mathrm{Std}(W)\cdot\mathrm{Std}(Z)} = -\frac{\frac{1}{4}}{\sqrt{\frac{23}{36}}\cdot\sqrt{\frac{1}{4}}} = -\sqrt{\frac{3}{23}} \approx -0.361.$$

Example 3.3.6 (a continuous 2D distribution with nontrivial correlation). A random vector (X,Y) has a continuous joint p.d.f. of the form

$$f_{(X,Y)}(x,y) = \begin{cases} C(1-(x+y)) & \text{for } x,y \geq 0,\ x+y \leq 1; \\ 0 & \text{elsewhere.} \end{cases}$$

The constant C can be determined from the normalization condition,

$$\int_0^1 \int_0^{1-x} C(1-(x+y))\,dy\,dx = 1,$$

which gives $C = 6$. The plot of the surface representing this density is given in Figure 3.3.2.

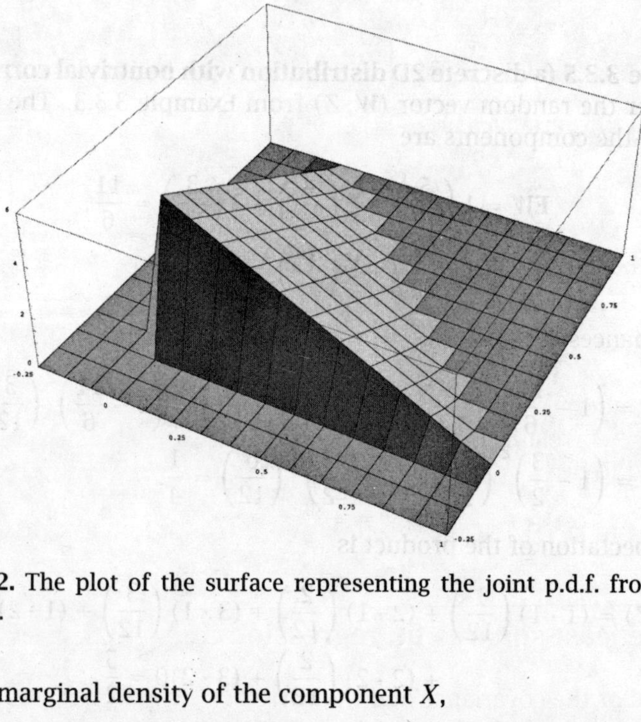

Fig. 3.3.2. The plot of the surface representing the joint p.d.f. from Example 3.3.6.

The marginal density of the component X,

$$f_X(x) = \int_0^{1-x} 6(1-(x+y))\,dy = 3(1-x)^2,$$

for $0 < x < 1$. It is equal to 0 elsewhere, and its plot is pictured in Figure 3.3.3.

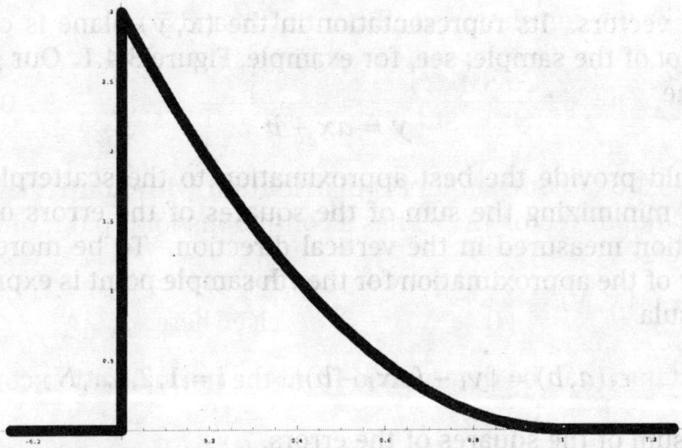

Fig. 3.3.3. The marginal density $F_X(x)$ of the X component of the random vector from Example 3.3.6.

The expectations of X and Y are easily evaluated using the marginal p.d.f.:

$$\mathrm{E}X = \mathrm{E}Y = \int_0^1 x \cdot 3(1-x)^2 dx = \frac{1}{4}.$$

Similarly, the variances are

$$\sigma^2(X) = \sigma^2(Y) = \int_0^1 \left(x - \frac{1}{4}\right)^2 \cdot 3(1-x)^2 dx = \frac{3}{80}.$$

Finally, the covariance is

$$\mathrm{Cov}(X,Y) = \int_0^1 \int_0^{1-x} \left(x - \frac{1}{4}\right)\left(y - \frac{1}{4}\right) \cdot 6(1-(x+y))dydx = -\frac{1}{80}.$$

So the random components X and Y are not independent; they are negatively correlated. The correlation coefficient itself is now easily evaluated to be

$$\rho_{X,Y} = \frac{-\frac{1}{80}}{\frac{3}{80}} = -\frac{1}{3}.$$

3.4 The least-squares fit, regression line

The roles of the covariance and the correlation coefficient will become better understood in the context of the following *least-squares regression* problem.

Consider a sample,

$$(x_1, y_1), (x_2, y_2), \ldots, (x_N, y_N),$$

of N 2D vectors. Its representation in the (x, y)-plane is called the *scatterplot* of the sample; see, for example, Figure 3.4.1. Our goal is to find a line

$$y = ax + b$$

that would provide the best approximation to the scatterplot in the sense of minimizing the sum of the squares of the errors of the approximation measured in the vertical direction. To be more precise, the error of the approximation for the ith sample point is expressed by the formula

$$\epsilon_i(a,b) = |y_i - (ax_i - b)|, \qquad i = 1, 2, \dots, N,$$

and the sum of the squares of the errors,

$$\sum_{i=1}^{N} \epsilon_i^2(a,b) = \sum_{i=1}^{N} (y_i - (ax_i - b))^2,$$

is a nice, differentiable function of two variables a and b. We can find its minimum by taking partial derivatives with respect to a and b and equating them to 0:[20]

$$\frac{\partial}{\partial a} \sum_{i=1}^{N} \epsilon_i^2(a,b) = -2 \sum_{i=1}^{N} (y_i - (ax_i + b))x_i = 0,$$
$$\frac{\partial}{\partial b} \sum_{i=1}^{N} \epsilon_i^2(a,b) = -2 \sum_{i=1}^{N} (y_i - (ax_i + b)) = 0.$$

These two equations, sometimes called the *normal equations*, are linear in a and b and can be easily solved by the substitution method. To make the next step more transparent, we will introduce the following simplified notation for different sample means (think here about the means of random quantities with N possible values with each value assigned probability $\frac{1}{N}$):

$$\overline{x} = \frac{1}{N} \sum_{i=1}^{M} x_i, \qquad \overline{y} = \frac{1}{N} \sum_{i=1}^{M} y_i,$$
$$\overline{x^2} = \frac{1}{N} \sum_{i=1}^{M} x_i^2, \qquad \overline{y^2} = \frac{1}{N} \sum_{i=1}^{M} y_i^2,$$
$$\overline{xy} = \frac{1}{N} \sum_{i=1}^{M} x_i y_i.$$

[20] This explains why we consider quadratic errors rather than the straight absolute errors; in the latter case the calculus tools would not work so well.

Now the normal equations for a and b can be written in the form

$$a\overline{x} + b - \overline{y} = 0 \quad \text{and} \quad a\overline{x^2} + b\overline{x} - \overline{xy} = 0,$$

which can be immediately solved to give

$$b = \overline{y} - a\overline{x}, \quad a = \frac{\overline{xy} - \overline{x} \cdot \overline{y}}{\overline{x^2} - (\overline{x})^2}.$$

The first of the above two equations indicates that the point with coordinates formed by the sample means $\overline{x}$ and $\overline{y}$ is located on the regression line. To better see the meaning of the second equation, observe that

$$\overline{xy} - \overline{x} \cdot \overline{y} = \frac{1}{N}\sum_{i=1}^{N}(x_i - \overline{x})(y_i - \overline{y}) = \text{Cov}(x, y)$$

is just the sample covariance of the x- and y-coordinates, and that

$$\overline{x^2} - (\overline{x})^2 = \text{Var}(x), \qquad \overline{y^2} - (\overline{y})^2 = \text{Var}(y).$$

Thus the equation $y = ax + b$ of the regression line can now be rewritten in the elegant form

$$\frac{y - \overline{y}}{\sqrt{\text{Var}(y)}} = \rho_{x,y}\frac{x - \overline{x}}{\sqrt{\text{Var}(x)}}, \tag{3.4.1}$$

where

$$\rho_{x,y} = \frac{\text{Cov}(x, y)}{\sqrt{\text{Var}(x)}\sqrt{\text{Var}(y)}} = \frac{\text{Cov}(x, y)}{\sigma_x\sigma_y},$$

is the sample correlation coefficient. Its significance is now clear: $\rho_{x,y}$ is the slope of the regression line but only after the x- and y-coordinates were standardized (see (3.2.11)), that is, they were centered by the means $\overline{x}$ and $\overline{y}$, and rescaled by the standard deviations σ_x and σ_y, respectively.

Example 3.4.1. Consider a 2D vector sample of size 10 (see Table 3.4.1).

Table 3.4.1.

x	1.06	2.08	3.28	4.13	5.28	6.39	7.12	8.04	9.23	10.38
y	1.10	3.37	3.23	6.92	7.66	6.78	8.12	9.94	9.55	10.87

The coefficients are $a = 0.9934$ and $b = 1.0925$, so that the equation of the regression line is

$$y = 0.9934 \cdot x + 1.0925$$

and the correlation coefficient is

$$\rho_{x,y} = 0.1063.$$

The scatterplot of these data as well as the plot of the regression line are shown in Figure 3.4.1.

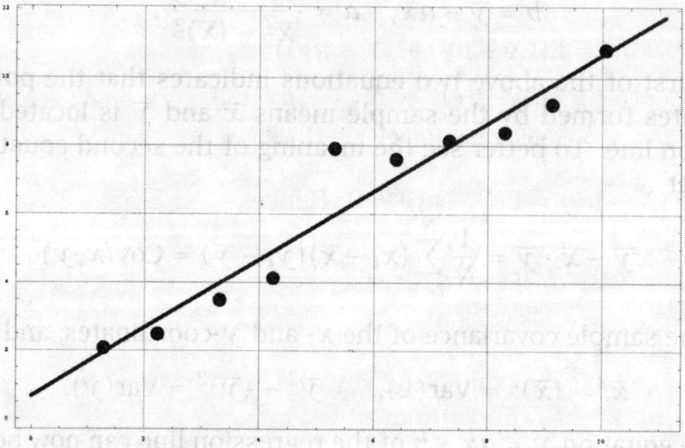

Fig. 3.4.1. The scatterplot and the least-squares fit regression line for data from Example 3.4.1.

3.5 The law of large numbers and the stability of fluctuations law

One of the fundamental theorems of statistics, called *the law of large numbers* (*LLN*), says that if $X_1, X_2, \ldots, X_n$ are independent random quantities with identical probability distributions (i.i.d.) and finite identical expectations $\mathbf{E}X_i = \mu_X$, then as $n \to \infty$, the averages converge to that expectation, i.e.,

$$\overline{X}_n \equiv \frac{X_1 + X_2 + \cdots + X_n}{n} \longrightarrow \mu_X \quad \text{as } n \to \infty. \tag{3.5.1}$$

Of course, the immediate issue is what do we mean by the convergence of random variables $\overline{X}_n$. For the purposes of these lectures the convergence of $\overline{X}_n$ to μ_X will mean that the standard deviation of the fluctuations of the averages $\overline{X}_n$ around the mean μ_X, that is, the differences $\overline{X}_n - \mu_X$, converge to zero as $n \to \infty$. More formally,

$$\lim_{n\to\infty} \operatorname{Std}(\overline{X}_n - \mu_X) = 0. \tag{3.5.2}$$

The statement (3.5.2) can be easily verified if we observe first that, for independent random quantities X and Y with finite variances, the variance

$$\mathrm{Var}(X+Y) = \mathrm{Var}(X) + \mathrm{Var}(Y), \tag{3.5.3}$$

which follows immediately from the multiplicative property (3.3.23) of the expectations of independent random variables; see Section 3.3. Indeed, if X and Y are independent, then $X - \mu_X$ and $Y - \mu_Y$ are also independent, so that

$$\begin{aligned}\mathrm{Var}(X+Y) &= \mathbf{E}((X-\mu_X)+(Y-\mu_Y))^2 \\ &= \mathbf{E}(X-\mu_X)^2 + 2\mathbf{E}(X-\mu_X)\mathbf{E}(Y-\mu_Y) + \mathbf{E}(Y-\mu_Y)^2 \\ &= \mathrm{Var}(X) + \mathrm{Var}(Y),\end{aligned}$$

because $\mathbf{E}(X-\mu_X) = \mathbf{E}(Y-\mu_Y) = 0$. Hence

$$\mathrm{Var}(\overline{X}_n - \mu_X) = \mathrm{Var}\left(\frac{X_1-\mu_X}{n} + \cdots + \frac{X_n-\mu_X}{n}\right) = \frac{\mathrm{Var}(X)}{n}, \tag{3.5.4}$$

which obviously approaches 0 as $n \to \infty$. Thus the *law of large numbers* (3.5.1), also often called the law of averages, is verified, at least in the situation when random quantities X_i have well-defined finite variances.[21]

A more subtle insight about the averages is provided by the following *stability of fluctuations law*, usually called the *central limit theorem* (*CLT*) in the mathematical and statistical literature. It states that as the averages $\overline{X}_n$ fluctuate around the expectation μ_X, the fluctuations, if viewed under a "magnifying glass," turn out to follow, asymptotically as $n \to \infty$, a Gaussian or normal probability distribution. More precisely, the c.d.f. of the standardized (see (3.2.13) and (3.5.4)) random fluctuations of the averages $\overline{X}_n$ around the mean μ_X,

$$Z_n = \frac{\sqrt{n}}{\mathrm{Std}(X)} \cdot (\overline{X}_n - \mu_X), \tag{3.5.5}$$

converges to the standard $N(0,1)$ Gaussian c.d.f., that is,

$$\lim_{n\to\infty} \mathbf{P}(Z_n \le z) = \Phi(z) \equiv \int_{-\infty}^{z} \phi(x)\,dx, \tag{3.5.6}$$

where the density is

$$\phi(z) = \frac{1}{\sqrt{2\pi}} e^{-z^2/2}. \tag{3.5.7}$$

The important assumption of the central limit theorem is that the common variance of X_is is finite. It can be immediately verified that all of Z_ns have mean zero and variance one; see (3.2.13) and (3.5.4), but the

[21] Note that not all random quantities have well-defined, finite variances; see Section 3.7.

proof of the convergence to a Gaussian limit is more delicate; for its sketch, see Section 3.7.

So the central limit theorem can be loosely rephrased as follows: *Standardized random fluctuations of averages of independent and identically distributed random quantities around their common expected value have a limiting standard Gaussian p.d.f.*

3.6 Estimators of parameters and their accuracy; confidence intervals

The law of large numbers can be reinterpreted as follows: If $X_1, X_2, \ldots, X_n$ is an i.i.d. sample from a certain probability distribution $F_X(x)$, then as n increases, the sample means $\overline{X}_n$, $n = 1, 2, \ldots$, become better and better estimators for the expectation of that distribution. In the statistical terminology the law of large numbers (3.5.1) says that $\overline{X}_n$ is a *consistent estimator* for parameter μ_X.

The central limit theorem (3.5.5)–(3.5.7) permits us to say what is the error of approximation of the theoretical mean μ_X by the sample mean $\overline{X}_n$, or, in other words, to establish the accuracy of the above estimation. Indeed, for a given sample of size n, the CLT says that the difference between the parameter μ_X and its estimator, the sample mean $\overline{X}_n$, is, after normalization by $\frac{\sqrt{n}}{\mathrm{Std}(X)}$, approximately $N(0,1)$ distributed so that, for large n,

$$\mathbf{P}\left(-\epsilon\frac{\mathrm{Std}(X)}{\sqrt{n}} \le \overline{X}_n - \mu_X \le \epsilon\frac{\mathrm{Std}(X)}{\sqrt{n}}\right) \approx \Phi(\epsilon) - \Phi(-\epsilon) \tag{3.6.1}$$
$$= 2\Phi(\epsilon) - 1 = C,$$

where $C = C(\epsilon)$ is a nonrandom constant, depending on the choice of ϵ only.

If X itself has a Gaussian p.d.f., then the above approximate equality becomes exact for all n. This follows from the fact that the sum of two independent Gaussian random quantities is again a Gaussian random quantity, obviously with the mean and variance being the sums of means and variances, respectively, of the corresponding random summands; see Section 3.7.

The above statement can be reformulated as follows: *the parameter μ_X is contained in the random interval*

$$\left(\overline{X}_n - \epsilon\frac{\mathrm{Std}(X)}{\sqrt{n}}, \overline{X}_n + \epsilon\frac{\mathrm{Std}(X)}{\sqrt{n}}\right)$$

---

with probability C. This statement is sometimes abbreviated by writing

$$\mu_X = \overline{X}_n \pm \epsilon \frac{\text{Std}(X)}{\sqrt{n}}$$

at the confidence level C. Note that it is the center of the above random interval that is random; its length is not random unless $\text{Std}(X) = \sigma_X$ itself has to be estimated from the sample.

Example 3.6.1 (a 95% confidence interval for μ_X with known σ_X). One hundred independently repeated measurements of a random quantity X were conducted, resulting in $\overline{X}_{100} = 7.1$. Suppose that we know that $\text{Std}(X) = 0.5$. To find the 95% confidence interval for μ_X using (3.5.1), we need to find ϵ such that $2\Phi(\epsilon) - 1 = 0.95$. From the table of the Gaussian $N(0,1)$ c.d.f. we have $z = 1.96$. Thus at the 95% confidence level,

$$7.1 - 1.96\frac{0.5}{\sqrt{100}} \le \mu_X \le 7.1 + 1.96\frac{0.5}{\sqrt{100}},$$

that is, $\mu_X = 7.1 \pm 0.098$ at the 95% confidence level. The above approximate confidence interval is exact if X has a Gaussian distribution.

Remark (error of the Gaussian approximation in the CLT). To be honest, we left open the essential, but delicate question of how good is the approximate equality in the basic formula (3.5.1) or, equivalently, the question of precise estimation of the error in the central limit theorem (3.5.6), which just says that the difference

$$\mathbf{P}(Z_n \le z) - \Phi(z) \to 0 \quad \text{as } n \to \infty,$$

where

$$Z_n = \frac{(X_1 + \cdots + X_n) - n\mu_X}{\sqrt{n} \cdot \text{Std}(X)}$$

are standardized sums $X_1 + \cdots + X_n$. It turns out that the accuracy in CLT is actually pretty good if the X_is have finite higher absolute moments. In particular, if the third central moment $M_3 = \mathbf{E}|X - \mu_X|^3 < \infty$, then, for all $-\infty < x < \infty$ and $n = 1, 2, \ldots$,

$$|\mathbf{P}(Z_n \le z) - \Phi(z)| \le \frac{kM_3}{\sqrt{n}\sigma_X^3},$$

where k is a universal (independent of n and X) constant contained in the interval $(0.4097, 0.7975)$. Its exact value is not known.[22]

Of course, the above procedure used in Example 3.4.1 requires advance knowledge of the standard deviation $\text{Std}(X)$. If that parameter is

[22] This error estimate in the CLT is known as the Berry-Esseen theorem and its proof can be found, for example, in V. V. Petrov's monograph *Sums of Independent Random Variables*, Springer-Verlag, Berlin, 1975.

unknown, then the obvious step is to try to estimate it from the sample $X_1, X_2, \ldots, X_n$ itself using the sample variance estimator

$$S_n^2 = \frac{1}{n-1}\sum_{i=1}^{n}(X_i - \overline{X})^2. \tag{3.6.2}$$

But in this case, even in the case of Gaussian X_i, the random quantity

$$T = \frac{\sqrt{n}}{S_n}(\overline{X} - \mu_X) \tag{3.6.3}$$

is no longer $N(0,1)$ distributed, so a simple construction of the confidence interval for μ_X is no longer possible.

However, in the case of a Gaussian random sample $X_1, X_2, \ldots, X_n$, it is known[23] that the random quantity T has the p.d.f.

$$f_T(x; n-1) = \frac{\Gamma\left(\frac{n}{2}\right)}{\sqrt{n\pi}\,\Gamma\left(\frac{n-1}{2}\right)}\left(1 + \frac{x^2}{n-1}\right)^{-n/2}, \tag{3.6.4}$$

which traditionally is called the *Student-T* p.d.f. with $(n-1)$ degrees of freedom. The Gamma function $\Gamma(\alpha)$ appearing in the definition of f_T is defined by the formula

$$\Gamma(\alpha) = \int_0^\infty x^{\alpha-1}e^{-x}\,dx, \quad \alpha > 0. \tag{3.6.5}$$

It is worth noting that

$$\alpha\Gamma(\alpha) = \Gamma(\alpha+1) \quad \text{and} \quad \Gamma(n) = (n-1)!, \tag{3.6.6}$$

if n is a positive integer. Thus the Gamma function is an interesting extension of the concept of the factorial to noninteger numbers.

Therefore, in this case, the $C = (2F_T(\epsilon) - 1)$ confidence interval for μ_X is of the form

$$\left(\overline{X}_n - \epsilon\frac{S_n}{\sqrt{n}}, \overline{X}_n + \epsilon\frac{S_n}{\sqrt{n}}\right). \tag{3.6.7}$$

It is convenient to tabulate the quantiles $Q_T(\alpha; n)$ defined by the condition

$$F_T(Q_T(\alpha; n)) = \alpha, \tag{3.6.8}$$

[23] See, for example, M. Denker and W. A. Woyczyński, *Introductory Statistics and Random Phenomena: Uncertainty, Complexity, and Chaotic Behavior in Engineering and Science*, Birkhäuser Boston, Cambridge, MA, 1998, for more details on the statistical issues discussed in this section.

rather than the c.d.f. itself. Note that the quantile is just the function inverse to c.d.f. The tables of selected quantiles $Q_T(\alpha; n)$ are provided at the end of the chapter. Using the T-quantiles allows the C confidence level interval for μ_X to be simply written in the form

$$\left(\overline{X}_n - Q_T(1 + C2; n-1)\frac{S_n}{\sqrt{n}}, \overline{X}_n + Q_T\left(\frac{1+C}{2}; n-1\right)\frac{S_n}{\sqrt{n}}\right). \quad (3.6.9)$$

For large n, say, $n > 20$, the Student-T p.d.f. with n degress of freedom becomes almost indistinguishable from the $N(0,1)$ p.d.f. (see Exercise 3.7.17), and the latter can be used in the construction of confidence intervals even in the case of unknown variance.

Example 3.6.2 (a 90% confidence interval for μ_X with unknown Std(X)). Nine independent measurements of a Gaussian random quantity X resulted in $\overline{X}_9 = 2.56$ and $S_9 = 0.12$. With the desired confidence level $C = 0.9$, the table yields the quantile

$$Q_T\left(\frac{1+0.9}{2}; 8\right) = Q_T(0.95; 8) = 1.86.$$

Hence the 90% confidence interval for the expectation μ_X is of the form

$$\left(2.56 - 1.86 \cdot \frac{0.12}{\sqrt{9}}, 2.56 + 1.86 \cdot \frac{0.12}{\sqrt{9}}\right) = (2.56 - 0.07, 2.56 + 0.07)$$

or, in other words, $\mu_X = 2.56 \pm 0.07$ at the 90% confidence level.

The final question in this section is, how good is the sample variance estimator S_n^2 introduced in (3.6.2)? Here again, the answer is difficult for a general c.d.f. F_X. However, in the case of a Gaussian sample one can prove that the nonnegative random quantity

$$\chi^2 = \frac{1}{\sigma_X^2}\sum_{i=1}^{n}(X_i - \overline{X}_n)^2 \quad (3.6.10)$$

has a p.d.f. of the form

$$f_{\chi^2}(x; n-1) = \frac{1}{2^{(n-1)/2}}\Gamma\left(\frac{n-1}{2}\right)x^{(n-3)/2}e^{-x/2}, \quad x \geq 0, \quad (3.6.11)$$

which traditionally is called the chi-square p.d.f. with $(n-1)$ degrees of freedom.[24] Thus the C confidence level interval for σ_X^2 is of the form

$$\left(\frac{(n-1)S_X^2}{Q_{\chi^2}\left(\frac{1+C}{2}; n-1\right)}, \frac{(n-1)S_X^2}{Q_{\chi^2}\left(\frac{1-C}{2}; n-1\right)}\right) \quad (3.6.12)$$

Selected quantiles $Q_{\chi^2}(\alpha; n)$ for the chi-square distributions are given in the tables at the end of this chapter.

[24] Compare this definition with the calculation of the p.d.f. of the square of the $N(0,1)$-distributed random quantity in Example 3.1.11.

Example 3.6.3 (a 99% confidence interval for Var(X)). Twenty-six independent measurements of a Gaussian random quantity X resulted in the estimate $S_{26}^2 = 1.37$ for the variance Var(X). With $C = 0.99$, the tables yield

$$Q_{\chi^2}\left(\frac{1+0.99}{2};25\right) = Q_{\chi^2}(0.995;25) = 46.928$$

and

$$Q_{\chi^2}\left(\frac{1-0.99}{2};25\right) = Q_{\chi^2}(0.005;25) = 10.520.$$

Thus the 99% confidence level interval for the variance σ_X^2 is

$$\left(\frac{25\cdot 1.37}{46.928}, \frac{25\cdot 1.37}{10.520}\right) = (0.723, 3.255).$$

The interval is relatively large because the confidence level demanded is very high. Note that it is not symmetric about the estimated value $S_{26}^2 = 1.37$.

3.7 Problems, exercises, and tables

Use *Mathematica*, MAPLE, or MATLAB as needed throughout this and other problem sections.

3.7.1. Plot the c.d.f.s of binomial quantities X with $p = 0.21$ and $n = 5, 13, 25$. Calculate the probabilities that X takes values between 1.3 and 3.7. Repeat the same exercise for $p = 0.5$ and $p = 0.9$.

3.7.2. Calculate the probability that a random quantity uniformly distributed over the interval $[0, 3]$ takes values between 1 and 3. Do the same calculation for the exponentially distributed random quantity with parameter $\mu = 1.5$, and the Gaussian random quantity with parameters $\mu = 1.5$, $\sigma^2 = 1$.

3.7.3. Prove that $\alpha\Gamma(\alpha) = \Gamma(\alpha + 1)$, and that $\Gamma(n) = (n-1)!$. Use the integration-by-parts formula. Verify analytically that $\Gamma(\frac{1}{2}) = \sqrt{\pi}$. Use the idea employed in Example 3.1.6 to prove that the standard Gaussian density is normalized.

3.7.4. The p.d.f. of a random variable X is expressed by the quadratic function $f_X(x) = ax(1-x)$, for $0 < x < 1$, and is zero outside the unit interval. Find a from the normalization condition and then calculate $F_X(x)$, $\mathbf{E}X$, Var(X), Std(X), the nth central moment, and $\mathbf{P}(0.4 < X < 0.9)$. Graph $f_X(x)$ and $F_X(x)$.

3.7.5. Find the c.d.f and p.d.f. of the random quantity $Y = X^3$, where X is uniformly distributed on the interval $[1, 3]$.

3.7.6. Find the c.d.f and p.d.f. of the random quantity $Y = \tan X$, where X is uniformly distributed over the interval $(-\frac{\pi}{2}, \frac{\pi}{2})$. Find a physical (geometric) interpretation of this result.

3.7.7. Verify that $\mathrm{Var}(X) = \mathbf{E}X^2 - (\mathbf{E}X)^2$; see formula (3.2.6).

3.7.8. Calculate the expectation and the variance of the binomial distribution from Example 3.1.2.

3.7.9. Calculate the expectation and the variance of the Poisson distribution from Example 3.1.3.

3.7.10. Calculate the expectation, the variance, and the nth moment of the exponential distribution from Example 3.1.5.

3.7.11. Calculate the nth central moment of the Gaussian distribution from Example 3.1.6.

3.7.12. Derive the formula for the binomial distribution from Example 3.1.2, relying on the observation that it is the distribution of the sum of n independent and identically distributed Bernoulli random qunatities. Show that if $p = \frac{\mu}{n}$ and $n \to \infty$, then the binomial probabilities converge to the Poisson probabilities.

3.7.13. A random quantity X has an even p.d.f. $f_X(x)$ of the triangular shape shown in Figure 3.7.1.

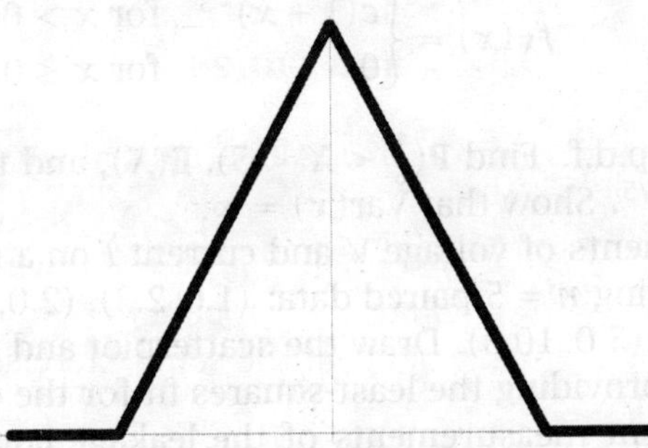

Fig. 3.7.1.

(a) How many parameters do you need to describe this p.d.f.? Find an explicit analytic formula for p.d.f. $f_X(x)$ and c.d.f. $F_X(x)$. Graph both of them.

(b) Find the expectation and variance of X.

(c) Let $Y = X^3$. Find the p.d.f. $f_Y(y)$ and graph it.

3.7.14. A discrete 2D random vector (X, Y) has the following joint p.d.f.:

$$\mathbf{P}(X = 1, Y = 1) = \frac{2}{12}, \qquad \mathbf{P}(X = 2, Y = 1) = \frac{1}{12},$$

$$\mathbf{P}(X = 3, Y = 1) = \frac{1}{12}, \qquad \mathbf{P}(X = 1, Y = 3) = \frac{2}{12},$$

$$\mathbf{P}(X = 2, Y = 3) = \frac{4}{12}, \qquad \mathbf{P}(X = 3, Y = 2) = \frac{2}{12}.$$

Find the marginal distributions of X and Y, their expectations and variances, as well as the covariance and the correlation coefficient of X and Y. Are X and Y independent?

3.7.15. Verify the Cauchy-Schwartz inequality (3.3.19). *Hint:* Take $Z = \frac{X-\mathbf{E}X}{\sigma(X)}$ and $W = \frac{Y-\mathbf{E}Y}{\sigma(Y)}$, and consider the discriminant of the expression $\mathbf{E}(Z + xW)^2$. The latter is quadratic in variable x and necessarily always nonnegative, so it can have at most one root.

3.7.16. The following sample of random vector (X, Y) was obtained: $(1, 1.7), (2, 2), (5, 4.3), (7, 5.9), (9, 8), (9, 8.7)$. Produce the scatterplot of the sample and the corresponding least-squares regression line.

3.7.17. Using the table of $N(0, 1)$ c.d.f. provided at the end of this chapter calculate $\mathbf{P}(-1 \le Y \le 2)$ if $Y \sim N(0.7, 4)$.

3.7.18. Produce graphs of the Student-T p.d.f. $f_T(x, n)$, for $n = 2, 5, 12, 20$, and compare them with the standard normal p.d.f.

3.7.19. Produce graphs of the chi-square p.d.f. $f_{\chi^2}(x, n)$ for $n = 2, 5, 12, 20$.

3.7.20. Find a constant $c > 0$ such that the function

$$f_X(x) = \begin{cases} c(1+x)^{-3} & \text{for } x > 0; \\ 0 & \text{for } x \le 0. \end{cases}$$

is a valid p.d.f. Find $\mathbf{P}(\frac{1}{5} < X < 5)$, $\mathbf{E}(X)$, and the p.d.f. $f_Y(y)$ of $Y = X^{1/5}$. Show that $\mathrm{Var}(x) = \infty$.

3.7.22. Measurements of voltage V and current I on a resistor yielded the following $n = 5$ paired data: $(1.0, 2.3), (2.0, 4.1), (3.0, 6.4), (4.0, 8.5), (5.0, 10.5)$. Draw the scatterplot and find the regression line providing the least-squares fit for the data.

3.7.23. Independent measurements of the leakage current I on a capacitor yielded the following data: 2.71, 2.66, 2.78, 2.67, 2.71, 2.69, 2.70, 2.73 mA. Assuming that the distribution of the random quantity I is Gaussian, find the 95% confidence intervals for the expectation $\mathbf{E}I$ and the variance σ_X^2.

3.7.24. Complete the following sketch of the proof of the central limit theorem.

(a) Define $L_X(u)$ as the Laplace transform of c.d.f. $F_X(x)$:

$$L_X(u) = \int_{\infty}^{\infty} e^{ux}\, dF_X(x).$$

Find $L'_X(0)$ and $L''_X(0)$.

(b) Calculate $L_X(u)$ for the Gaussian $N(0, 1)$ random quantity.

(c) Prove that, for independent random quantities X and Y,

$$L_{X+Y}(u) = L_X(u) \cdot L_Y(u).$$

(d) Utilizing (c), calculate

$$L_{\sqrt{n}(\bar{X}-\mu_X)/\,\mathrm{Std}(X)}(u),$$

(it is easier to work here with the logarithm of the Laplace transform) and find its limit as $n \to \infty$. Compare it with the Laplace transform of the Gaussian $N(0,1)$ random quantity.

3.7.25. Use the introduced above Laplace transform technique to prove that the sum of two independent Gaussian random quantities is again a Gaussian random quantity.

3.7.26. What is the probability P that a randomly selected chord is shorter than the side S of an equilateral triangle inscribed in the circle? Here are two seemingly reasonable solutions:[25]

(a) A chord is determined by its two endpoints. Fix one of them to be A. For the chord to be shorter than the side S, the other endpoint must be chosen on either the arc AB or on the arc CA, and each of them is subtended by an engle of 120°. Thus $P = \frac{2}{3}$.

(b) A chord is completely determined by its center. For the chord to be shorter than the side S, the center must lie outside the circle of radius equal to the half of the radius of the original circle and the same center. Hence, the probability P equals the ratio of the annular area between two circles and the area of the original circle, which is $\frac{3}{4}$.

These two solutions are different. How is that possible?

3.7.27. Derive formulas for the c.d.f. $F_Y(y)$, and the p.d.f. $f_Y(y)$, of a transformation $Y = g(X)$ of a random quantity X, in terms of its c.d.f. $F_X(x)$, and p.d.f. $f_X(x)$, in the case when the transforming function $y = g(x)$ is *monotonically decreasing*. Follow the line of reasoning used to derive the analogous formulas (3.1.11)–(3.1.12) for *monotonically increasing* transformations. How would you extend these formulas to transformations that are monotonically increasing on some intervals and decreasing on their complement?

3.7.28. Verify that the components X, Y of the random vector with probability distribution $P((X,Y) = (1,0)) = P((X,Y) = (0,1)) = P((X,Y) = (-1,0)) = P((X,Y) = (0,-1)) = \frac{1}{4}$ are uncorrelated but not statistically independent. Calculate probability distribution of a random vector (W,Z) with statistically independent components and the same marginal distribution as (X,Y).

---

[25] For more information, see M. Denker and W. A. Woyczyński, *Introductory Statistics and Random Phenomena: Uncertainty, Complexity, and Chaotic Behavior in Engineering and Science*, Birkhäuser Boston, Cambridge, MA, 1998, Example 5.1.1.

Table 3.7.1. Gaussian $N(0,1)$ c.d.f. $\Phi(z) = (2\pi)^{-1/2} \int_{-\infty}^{z} e^{-x^2/2} dx$.

z	.00	.01	.02	.03	.04	.05	.06	.07	.08	.09
0.0	.5000	.5040	.5080	.5120	.5160	.5199	.5239	.5279	.5319	.5359
0.1	.5395	.5438	.5478	.5517	.5557	.5596	.5636	.5675	.5714	.5753
0.2	.5793	.5832	.5871	.5910	.5948	.5987	.6026	.6064	.6103	.6141
0.3	.6179	.6217	.6255	.6296	.6331	.6366	.6406	.6443	.6480	.6517
0.4	.6554	.6591	.6628	.6664	.6700	.6736	.6772	.6808	.6884	.6879
0.5	.6915	.6956	.6985	.7019	.7054	.7088	.7123	.7157	.7190	.7224
0.6	.7257	.7291	.7324	.7857	.7389	.7422	.7454	.7486	7517	.7549
0.7	.7580	.7611	.7642	7673	.7704	.7734	.7764	.7794	.7823	.7852
0.8	.7881	.7910	.7939	.7967	.7995	.8023	.8051	.8075	.8106	.8133
0.9	.8195	.8186	.8212	.8238	.8264	.8289	.8315	.8340	.8365	.8389
1.0	.8413	.8438	.8461	.8485	.8503	.8531	.8554	.8577	.8599	.8621
1.1	.8613	.8665	.8686	.8708	.8729	.8749	.8770	.8796	.8810	.8830
1.2	.8849	.8869	.8888	.8907	.8925	.8944	.8962	.8980	.8977	.9015
1.3	.9032	.9049	.9066	.9082	.9099	.9115	.9131	.9147	.9162	.9177
1.4	.9192	.9207	.9222	.9236	.9251	.9265	.9279	.9292	.9306	.9319
1.5	.9332	.9345	.9359	.9370	.9382	.9309	.9404	.9418	.9429	.9441
1.6	.9452	.9463	.9474	.9484	.9495	.9505	.9515	.9525	.9535	.9545
1.7	.9554	.9564	.9573	.9582	.9591	.9599	.9606	.9616	.9625	.9633
1.8	.9641	.9649	.9656	.9664	.9671	9678.	9666.	9693.	9699.	.9706
1.9	.9713	.9719	.9726	.9732	.9738	.9744	.9750	.9756	.9761	.9767
2.0	.9773	.9778	.9783	.9788	.9793	.9798	.9803	.9808	.9812	.9817
2.1	.9821	.9826	.9830	.9834	.9838	.9842	.9846	.9850.	9854	.9857
2.2	.9891	.9861	.9868	.9871	.9875	.9878	.9881	.9884	.9887	.9890
2.3	.9893	.9896	.9868	.9871	.9875	.9878	.9881	.9884	.9887	.9890
2.4	.9918	.9820	.9922	.9925	.9927	.9929	.9931	.9932	.9934	.9936
2.5	.9938	.9940	.9941	.9943	.9945	.9946	.9948	.9949	9951	.9952
2.6	.9953	.9955	.9956	.9957	.9959	.9960	.9961	.9962	.9963	.9964
2.7	.9965	.9966	.9967	.9968	.9969	.9970	.9971	.9972	.9973	.9974
2.8	.9974	.9975	.9976	.9977	.9977	.9978	.9979	.9979	.9980	.9981
2.9	.9981	.9982	.9983	.9983	.9984	.9984	.9985	.9985	.9986	.9986
3.0	.9987	.8876	.8876	.8877	.8877	.8878	.8878	.9989	.9990	.9990
3.1	.9990	.9991	.9991	.9991	.9992	.9992	.9992	.9992.	.9993	.9993
3.2	.9993	.9993	.9994	.9994	.9994	.9994	.9994	.9995	.9995	.9995
3.3	.9995	.9995	.9996	.9996	.9996	.9996	.9996	.9996	.9996	.9997
3.4	.9997	.9997	.9997	.9997	.9997	.9997	.9997	.9997	.9997	.9998

Table 3.7.2. Student-T distribution quantiles $Q_T(\alpha; n)$.

$n \backslash \alpha$	0.1000	0.0500	0.0250	0.0100	0.0050	0.0010	0.0005
1	3.078	6.314	12.706	31.821	63.657	318.317	636.61
2	1.886	2.920	4.303	6.965	9.925	22.326	31.598
3	1.638	2.353	3.182	4.451	5.841	10.213	12.924
4	1.533	2.132	2.776	3.747	4.604	7.173	8.610
5	1.476	2.015	2.571	3.365	4.032	5.893	8.610
6	1.440	1.943	2.447	3.143	3.707	5.208	5.959
7	1.415	1.895	2.365	2.998	3.500	4.785	5.408
8	1.397	1.860	2.306	2.896	3.355	4.501	5.041
9	1.383	1.833	2.262	2.821	3.250	4.297	4.781
10	1.372	1.813	2.228	2.764	3.169	4.144	4.587
11	1.364	1.796	2.201	2.718	3.106	4.025	4.437
12	1.356	1.782	2.179	2.681	3.055	3.930	4.318
13	1.350	1.771	2.160	2.650	3.012	3.852	4.221
14	1.345	1.761	2.145	2.624	2.977	3.787	4.141
15	1.341	1.753	2.131	2.602	2.947	3.733	4.073
16	1.337	1.746	2.120	2.584	2.921	3.686	4.015
17	1.333	1.740	2.110	2.567	2.898	3.646	3.965
18	1.330	1.734	2.101	2.553	2.879	3.610	3.992
19	1.328	1.729	2.093	2.540	2.861	3.579	3.883
20	1.325	1.725	2.086	2.528	2.845	3.552	3.849
21	1.323	1.721	2.080	2.518	2.831	3.527	3.819
22	1.321	1.717	2.074	2.508	2.819	3.505	3.792
23	1.320	1.714	2.069	2.500	2.807	3.485	3.768
24	1.318	1.711	2.064	2.492	2.797	3.467	3.745
25	1.316	1.708	2.059	2.485	2.787	3.450	3.725
26	1.315	1.706	2.056	2.479	2.779	3.435	3.707
27	1.314	1.703	2.052	2.473	2.771	3.421	3.690
28	1.312	1.701	2.049	2.467	2.763	3.408	3.674
29	1.311	1.699	2.045	2.462	2.756	3.396	3.659
30	1.311	1.697	2.042	2.457	2.750	3.385	3.646
40	1.303	1.684	2.021	2.423	2.704	3.307	3.551
60	1.296	1.671	2.000	2.390	2.660	3.232	3.460
120	1.289	1.658	1.980	2.358	2.617	3.160	3.373
∞	1.282	1.645	1.960	2.326	2.576	3.090	3.291

Table 3.7.3. Chi-square distribution quantiles $Q_{\chi^2}(\alpha; n)$.

$n\backslash\alpha$	0.9950	0.9900	0.9750	0.9500	0.9000	0.1000	0.0500	0.0250	0.0100	0.0050
1	0.000	0.000	0.001	0.004	0.016	2.706	3.843	5.025	6.637	7.882
2	0.010	0.020	0.051	0.103	0.211	4.605	5.992	7.378	9.210	10.597
3	0.072	0.115	0.216	0.352	0.584	6.251	7.815	9.348	11.344	12.937
4	0.207	0.297	0.484	0.711	1.064	7.779	9.488	11.143	13.277	14.860
5	0.412	0.554	0.831	1.145	1.160	9.236	11.070	12.832	15.085	16.748
6	0.676	0.872	1.237	1.635	2.204	10.645	12.592	14.440	16.812	18.548
7	0.989	1.239	1.690	2.167	2.833	12.17	14.067	16.012	18.474	20.276
8	1.344	1.646	2.180	2.733	3.490	13.362	15.507	17.534	20.090	21.954
9	1.735	2.088	2.700	3.325	4.168	14.684	16.919	19.022	21.665	23.587
10	2.156	2.558	3.247	3.940	4.865	15.987	18.307	20.483	23.209	25.188
11	2.603	3.053	3.816	4.575	5.578	17.275	19.675	21.920	24.724	26.755
12	3.074	3.571	4.404	5.226	6.304	18.549	21.026	23.337	26.217	28.300
13	3.565	4.107	5.009	5.892	7.041	19.812	22.362	24.735	27.687	29.817
14	4.075	4.660	5.629	6.571	7.790	21.064	23.685	26.119	29.141	31.319
15	4.600	5.229	6.262	7.261	8.547	22.307	24.996	27.488	30.577	32.799
16	5.142	5.812	6.908	7.962	9.312	23.542	26.296	28.845	32.000	34.267
17	5.697	6.407	7.564	8.682	10.085	24.769	27.587	30.190	33.408	35.716
18	6.265	7.015	8.231	9.390	10.865	25.989	28.869	31.526	34.805	37.156
19	6.843	7.632	8.906	10.117	11.651	27.203	30.143	32.852	36.190	38.580
20	7.434	8.260	9.591	10.851	12.443	28.412	31.410	34.170	37.566	39.997
21	8.033	8.897	10.283	11.591	13.240	29.615	32.670	35.479	38.930	41.399
22	8.643	9.542	10.982	12.338	14.042	30.813	33.924	36.781	40.289	42.796
23	9.260	10.195	11.688	13.090	14.848	32.007	35.172	38.075	41.637	44.179
24	9.886	10.856	12.401	13.848	15.659	33.196	36.415	39.364	42.980	45.558
25	10.519	11.523	13.120	14.611	16.473	34.381	37.652	40.646	44.313	46.925
26	11.160	12.198	13.844	15.379	17.292	35.563	38.885	41.923	45.642	48.290
27	11.807	12.878	14.573	16.151	18.114	36.741	40.113	43.194	46.962	49.642
28	12.461	13.565	15.308	16.928	18.939	37.916	41.337	44.461	48.278	50.993
29	13.120	14.256	16.147	17.708	19.768	39.087	42.557	45.772	49.586	52.333
30	13.787	14.954	16.791	18.493	20.599	40.256	43.773	46.979	50.892	53.672
31	14.457	15.655	17.538	19.280	21.433	41.422	44.985	48.231	52.190	55.000
32	15.134	16.362	18.291	20.072	22.271	42.585	46.194	49.480	53.486	56.328
33	15.814	17.073	19.046	20.866	23.110	43.745	47.400	50.724	54.774	57.646
34	16.501	17.789	19.806	21.664	23.952	44.903	48.602	51.966	56.061	58.964
35	17.191	18.508	20.569	22.465	24.796	46.059	49.802	53.203	57.340	60.272
36	17.887	19.233	21.336	23.269	25.643	47.212	50.998	54.437	58.619	61.581
37	18.584	19.960	22.105	24.075	26.492	48.363	52.192	55.667	59.891	62.880
38	19.289	20.691	22.878	24.884	27.343	49.513	53.384	56.896	61.162	64.181
39	19.994	21.425	23.654	25.695	28.196	50.660	54.572	58.119	62.462	65.473
40	20.706	22.164	24.433	26.509	29.050	51.805	55.758	59.342	63.691	66.766

4

Stationary Signals

In this chapter we introduce basic concepts necessary to study the time-dependent dynamics of random phenomena. The latter will be modeled as a family of random quantities indexed by a parameter, interpreted in this book as time. The parameter may be either continuous or discrete. Depending on the context and the tradition followed by different authors, such families are called *random signals, stochastic processes*, or *random time series*. The emphasis here is on random dynamics which are *stationary*, that is governed by underlying statistical mechanisms that do not change in time, although, of course, particular realizations of such families will be functions that vary with time. Think here about the random signal produced by the proverbial repeated coin tossing; the outomes vary while the fundamental mechanics remains the same.

4.1 Stationarity, autocovariance, and autocorrelation

A *random* (or *stochastic*) *signal* is a time-dependent family of random quantities $X(t)$. Depending on the context, one can consider random signals on the positive time line $t \geq 0$, on the whole time line $-\infty < t < \infty$, or on a finite time interval $t_0 \leq t \leq t_1$. Also it is useful to be able to consider random vector signals and signals with discrete time $t = \dots, -2, -1, 0, 1, 2, \dots$.

In this book we will restrict our attention to signals that are statistically stationary, which means that at least some of their statistical characteristics do not change in time. Several choices are possible here:

First-order strictly stationary signals. In this case, the c.d.f. $F_{X(t)}(x)$ does not change in time (it is time-shift invariant), that is,

$$F_{X(t)}(x) = F_{X(t+\tau)}(x) \quad \text{for all } t, \tau, x. \tag{4.1.1}$$

Second-order strictly stationary signals. In this case, the joint 2D c.d.f. $F_{(X(t_1),X(t_2))}(x_1, x_2)$ does not change in time, that is,

$$F_{(X(t_1),X(t_2))}(x_1, x_2) = F_{(X(t_1+\tau),X(t_2+\tau))}(x_1, x_2) \quad \text{for all } t_1, t_2, \tau, x_1, x_2. \tag{4.1.2}$$

In a similar fashion one can define the *nth-order strict stationarity* of random signal $X(t)$ as the time-shift invariance of the nD joint c.d.f., that is, the requirement that

$$F_{(X(t_1),\ldots,X(t_n))}(x_1, \ldots, x_n) = F_{(X(t_1+\tau),\ldots,X(t_n+\tau))}(x_1, \ldots, x_n) \tag{4.1.3}$$

for all $t_1, \ldots, t_n, \tau, x_1, \ldots, x_n$.

Finally, a random signal $X(t)$ is said to be *strictly stationary* if, for each $n = 1, 2, \ldots$, it is nth-order strictly stationary.

Obviously, as n increases, verifying the nth-order stationarity gets more and more difficult, not to mention practical difficulties that arise with checking the full strict stationarity. For this reason, a more modest concept of *second-order weakly stationary signals* is useful. In this case the invariance property is demanded only of the moments of the signal up to order 2. More precisely, a signal $X(t)$ is said to be second-order weakly stationary if its expectations and covariances are time-shift invariant, that is, if

$$\mu_X(t) \equiv \mathrm{E}[X(t)] = \mathrm{E}[X(t+\tau)] \equiv \mu_X(t+\tau) \tag{4.1.4}$$

for all t, τ, and the *autocovariance function* is

$$\begin{aligned}\gamma_X(t_1, t_2) &\equiv \mathrm{Cov}(X(t_1), X(t_2)) \\ &= \mathrm{Cov}(X(t_1+\tau), X(t_2+\tau)) \equiv \gamma_X(t_1+\tau, t_2+\tau)\end{aligned} \tag{4.1.5}$$

for all t_1, t_2, τ.

It is a consequence of the above two conditions that, for any second-order weakly stationary signal,

$$\mu_X(t) = \mu_X = \text{constant}, \tag{4.1.6}$$

and the autocovariance function depends only on the time lag $\tau = t_2 - t_1$,

$$\gamma_X(t_1, t_2) = \gamma_X(t_1 - t_1, t_2 - t_1) = \gamma_X(0, t_2 - t_1), \tag{4.1.7}$$

so that, in particular,

$$\mathrm{Var}(X(t)) \equiv \sigma^2_{X(t)} = \mathrm{Cov}(X(t), X(t)) = \gamma_X(0,0) = \sigma^2_X = \text{constant}. \tag{4.1.8}$$

Thus all the first and second moments of the signal can be expressed in terms of just two characteristics, the signal's *mean value* μ_X and signal's autocovariance function

$$\gamma_X(t) := \gamma_X(0,t) = \mathbf{E}[(X(0) - \mu_X)(X(t) - \mu_X)], \tag{4.1.9}$$

which is, as a result of the stationarity assumption, a function of just a single variable.

In the remainder of this discussion, we will restrict our attention to second-order weakly stationary signals $X(t)$, which we will simply call *stationary signals*. We will analyze them assuming only the knowledge of their mean value μ_X and their autocovariance function $\gamma_X(t)$.

The following properties of the autocovariance function follow directly from its definition and the Schwartz inequality (see Section 3.7):

$$\gamma_X(-t) = \gamma_X(t), \tag{4.1.10}$$

and

$$|\gamma_X(t)| \le \gamma_X(0) = \sigma^2_X. \tag{4.1.11}$$

In other words, the covariance function is even and its absolute value is bounded by its value at $t = 0$, where it is simply equal to the signal's variance.

In different situations it is often convenient to use close relatives of the autocovariance function, such as the *autocorrelation function*[26]

$$\begin{aligned}\phi_X(t) &= \mathbf{E}(X(t_1)X(t_1+t))\\ &= \mathrm{Cov}(X(t_1), X(t_1+t)) + \mathbf{E}(X(t_1)) \cdot \mathbf{E}(X(t_1+t)) = \gamma_X(t) + \mu_X^2,\end{aligned} \tag{4.1.12}$$

and the *normalized autocovariance function*

[26] You may have noticed that in signal processing the traditional term "autocorrelation function" is at odds with the previously introduced term "correlation coefficient," which really corresponds to the above-introduced "normalized autocovariance function." But the terminology is so well established that we will stick with it.

$$\xi_X(\tau) = \frac{\gamma_X(t)}{\sigma_X^2} = \frac{\phi_X(t) - \mu_X^2}{\sigma_X^2} \tag{4.1.13}$$

which has the advantage of having its values always contained in the interval $[-1, 1]$.

If the signals' mean value is zero, then, of course, the autocovariance and the autocorrelation functions are identical:

$$\gamma_X(\tau) = \phi_X(\tau).$$

In what follows, unless explicitly stated otherwise, we will always assume that the signals under consideration have zero means so that the autocorrelation and the autocovariance are the same functions.

The reminder of this section is devoted to a series of examples of stationary data. The first, real-life example (see Figure 4.1.1) shows a sample of a 21-channel recording of the sleep electroencephalogram (EEG) of a neonate. The duration of this multidimensional random signal is one minute and the sampling rate is 64 Hz. This particular EEG was taken during the so-called mixed frequency sleep stage and, in addition to the EEG, it also shows related signals such as electrocardiogram (EKG), breathing signal, eye muscle contraction signal, etc. The signal's components seem stationary for some channels while other channels seem to violate the stationarity property. This can be due to some artifacts in the recordings caused, for example, by the physical movements of the infant or by the onset of a different sleep stage (active, passive, rapid eye movement (REM), etc.). The study of EEG signals provides important information on the state of the brain's neural network and, in the case of infants, can be used to assess the maturity level of their brains. In Section 4.2, we will provide a method to estimate the autocorrelation function for such real-life data.

Examples 4.1.1–4.1.6 provide various mathematical models of stationary signals. In those cases, the autocorrelation functions can be explicitely calculated.

Example 4.1.1 (a random harmonic oscillation). Consider a signal which is a simple harmonic oscillation with nonrandom frequency $f_0 = \frac{1}{P}$ but random amplitude A such that the second moment $\mathbf{E}A^2 < \infty$, and random phase Θ uniformly distributed over the period and independent of A. In other words,

$$X(t) = A\cos(2\pi f_0(t + \Theta)).$$

The signal is stationary because its mean value is

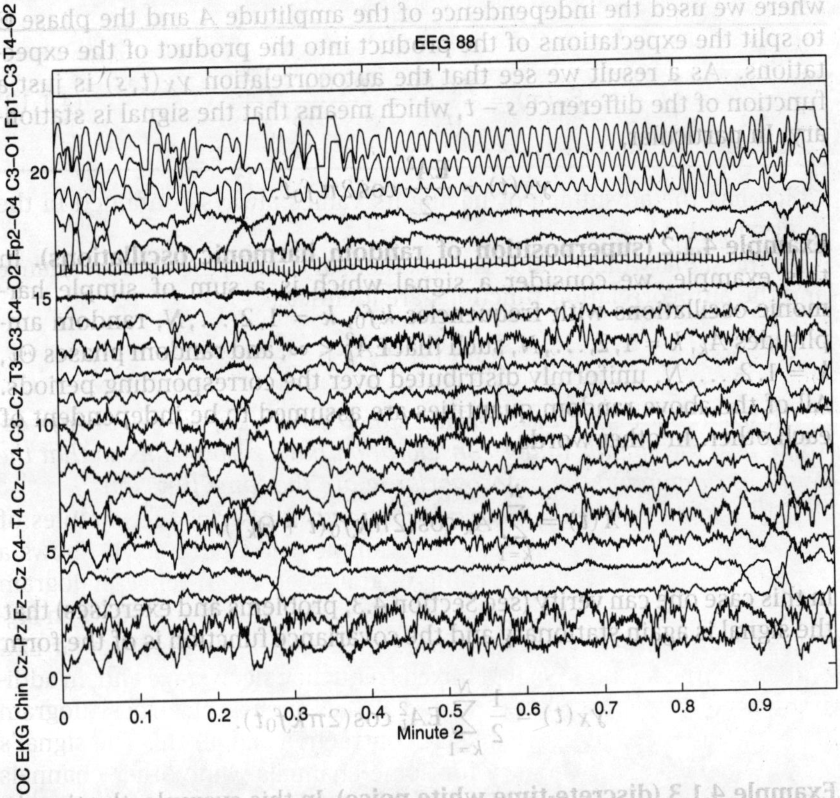

Fig. 4.1.1. A sample of a 21-channel recording of the sleep electroencephalogram (EEG) of a neonate. The duration of this multidimensional random signal is 60 seconds and the sampling rate is 64 Hz. (From A. Piryatinska's Ph.D. dissertation, Department of Statistics, Case Western Reserve University, Cleveland, 2004.)

$$\mathbf{E}X(t) = \mathbf{E}A\cos 2\pi f_0(t+\Theta) = \mathbf{E}A \cdot \int_0^P \cos 2\pi f_0(t+\theta)\,\frac{d\theta}{P} = \mathbf{E}A \cdot 0 = 0$$

and its autocovariance is

$$\begin{aligned}
\gamma_X(t,s) &= \mathbf{E}X(t)X(s) = \mathbf{E}[A\cos 2\pi f_0(t+\Theta)\cdot A\cos 2\pi f_0(s+\Theta)] \\
&= \mathbf{E}A^2 \cdot \int_0^P \cos 2\pi f_0(t+\theta)\cdot\cos 2\pi f_0(s+\theta)\,\frac{d\theta}{P} \\
&= \mathbf{E}A^2\frac{1}{2}\left(\int_0^P \cos 2\pi f_0(t+s+2\theta)\,\frac{d\theta}{P} + \int_0^P \cos 2\pi f_0(s-t)\,\frac{d\theta}{P}\right) \\
&= \frac{\mathbf{E}A^2}{2}\cos 2\pi f_0(s-t),
\end{aligned}$$

where we used the independence of the amplitude A and the phase Θ to split the expectations of the product into the product of the expectations. As a result we see that the autocorrelation $\gamma_X(t,s)$ is just a function of the difference $s-t$, which means that the signal is stationary. In particular,

$$\gamma_X(t) = \frac{\mathbf{E}A^2}{2}\cos 2\pi f_0 t.$$

Example 4.1.2 (superposition of random harmonic oscillations). In this example, we consider a signal which is a sum of simple harmonic oscillations with frequencies kf_0, $k = 1,2,\ldots,N$, random amplitudes A_k, $k = 1,2,\ldots,N$, such that $\mathbf{E}A_k^2 < \infty$, and random phases Θ_k, $k = 1,2,\ldots,N$, uniformly distributed over the corresponding periods. All of the above random quantities are assumed to be independent of each other. In other words,

$$X(t) = \sum_{k=1}^{N} A_k \cos(2\pi k f_0(t + \Theta_k)).$$

In this case one can verify (see Section 4.3, problems and exercises) that the signal is again stationary and the covariance function is of the form

$$\gamma_X(t) = \frac{1}{2}\sum_{k=1}^{N} \mathbf{E}A_k^2 \cos(2\pi k f_0 t).$$

Example 4.1.3 (discrete-time white noise). In this example, the time is discrete, that is, $t = n = \ldots,-2,-1,0,1,2,\ldots$ and the random signal $W(n)$ has mean zero and values at different times that are uncorrelated; its variance is σ_W^2. In other words,

$$\mu_W = 0,$$

and

$$\gamma_W(n,k) = \mathbf{E}(W(n)W(k)) = \begin{cases} \sigma_W^2 & \text{if } n-k=0, \\ 0 & \text{if } n-k \neq 0. \end{cases}$$

Note that the above-defined signal is stationary because its autocovariance (autocorrelation, since the mean is zero) is indeed a function of only the time lag and can be written in the form

$$\gamma_W(n,k) = \sigma_W^2 \delta(n-k),$$

where

$$\delta(n) = \begin{cases} 1 & \text{if } n = 0; \\ 0 & \text{if } n \neq 0, \end{cases}$$

is the discrete-time Dirac delta function. This kind of signal is called *discrete-time white noise* and it has mean zero and autocorrelation function

$$\gamma_W(n) = \sigma_W^2 \delta(n).$$

A sample path of a discrete-time white noise with $\sigma_W^2 = \frac{1}{12}$ is shown in Figure 4.1.2. It was produced using a random number generator in *Mathematica*, with the values of W_n uniformly distributed on the interval $[-\frac{1}{2}, \frac{1}{2}]$.

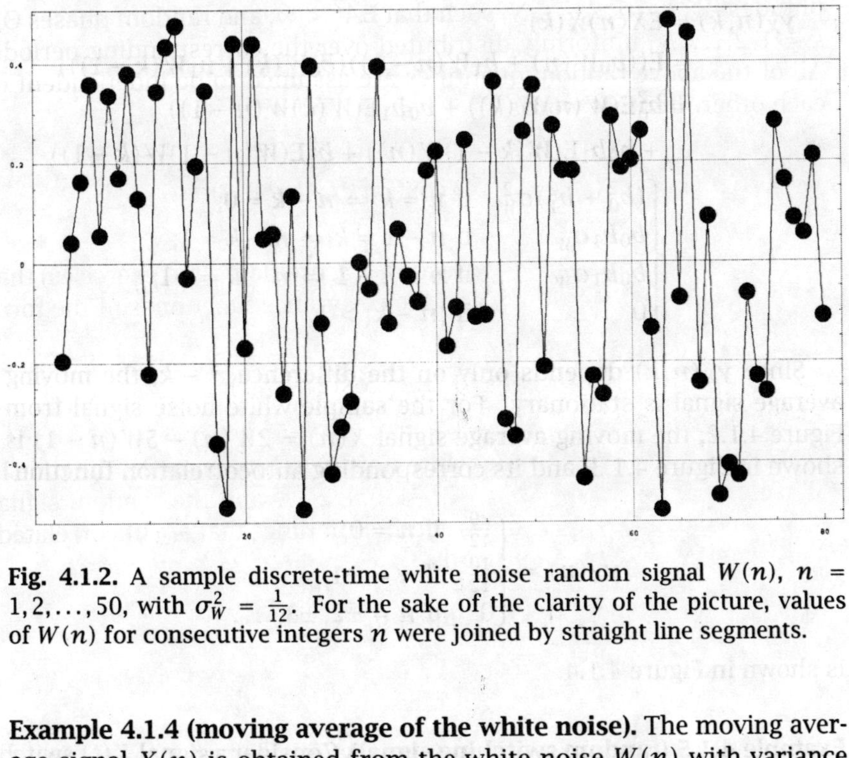

Fig. 4.1.2. A sample discrete-time white noise random signal $W(n)$, $n = 1, 2, \ldots, 50$, with $\sigma_W^2 = \frac{1}{12}$. For the sake of the clarity of the picture, values of $W(n)$ for consecutive integers n were joined by straight line segments.

Example 4.1.4 (moving average of the white noise). The moving average signal $X(n)$ is obtained from the white noise $W(n)$ with variance σ_W^2 by the "windowing" procedure. The windowing procedure mixes values of the white noise, $W(n), W(n-1), \ldots, W(n-q)$, in the time window of fixed width $q+1$, extending into the past, giving values with different time lags different weights, say, $b_0, b_1, \ldots, b_q$. More precisely,

$$X(n) = b_0 W(n) + b_1 W(n-1) + \cdots + b_q W(n-q).$$

You can interpret the moving average signal as a convolution of the white noise with the windowing weight sequence. One immediately obtains that $\mu_X = 0$. Since, for independent random quantities, the variance of the sum is equal to the sum of the variances, the variance is

$$\sigma_X^2 = \sigma_W^2 \sum_{i=0}^{q} b_i^2.$$

Calculation of the autocorrelation function is a little more complicated (see Section 4.3, problems and exercises) and here we will carry it out only in the case of the window of width 2, when

$$X(n) = b_0 W(n) + b_1 W(n-1).$$

Then

$$\begin{aligned}
\gamma_X(n,k) &= \mathbf{E}X(n)X(k) \\
&= \mathbf{E}((b_0W(n) + b_1W(n-1))(b_0W(k) + b_1W(k-1))) \\
&= b_0^2\mathbf{E}(W(n)W(k)) + b_0b_1\mathbf{E}(W(k)W(n-1)) \\
&\quad + b_0b_1\mathbf{E}(W(k-1)W(n)) + b_1^2\mathbf{E}(W(n-1)W(k-1)) \\
&= \begin{cases} (b_0^2 + b_1^2)\sigma_W^2 & \text{if } n = k \Leftrightarrow n - k = 0; \\ b_0b_1\sigma_W^2 & \text{if } n - 1 = k \Leftrightarrow n - k = 1; \\ b_0b_1\sigma_W^2 & \text{if } n = k - 1 \Leftrightarrow n - k = -1; \\ 0 & \text{if } |n-k| > 1. \end{cases}
\end{aligned}$$

Since $\gamma_X(n,k)$ depends only on the difference $n - k$, the moving average signal is stationary. For the sample white noise signal from Figure 4.1.2, the moving average signal $X(n) = 2W(n) + 5W(n-1)$ is shown in Figure 4.1.3, and its corresponding autocorrelation function

$$\gamma_X(n) = \begin{cases} \frac{29}{12} & \text{if } n = 0; \\ \frac{10}{12} & \text{if } n = \pm 1; \\ 0 & \text{if } n = \pm 2, \pm 3, \ldots \end{cases}$$

is shown in Figure 4.1.4.

Example 4.1.5 (random switching signal). Consider a signal $X(t)$ switching back and forth between values $+1$ and -1 at random times. More precisely, the intial value of the signal, $X(0)$, is a random quantity with the symmetric Bernoulli distribution, that is, $\mathbf{P}(X(0) = \pm 1) = \frac{1}{2}$, and the interswitching times form a sequence $T_1, T_2, \ldots$, of independent random quantities with the exponential distribution:

$$\mathbf{P}(T_i \le t) = 1 - e^{-t}, \quad t > 0,$$

of mean 1. The initial value $X(0)$ is assumed to be independent of interswitching times T_i. A typical sample of such a signal is shown in Figure 4.1.5.

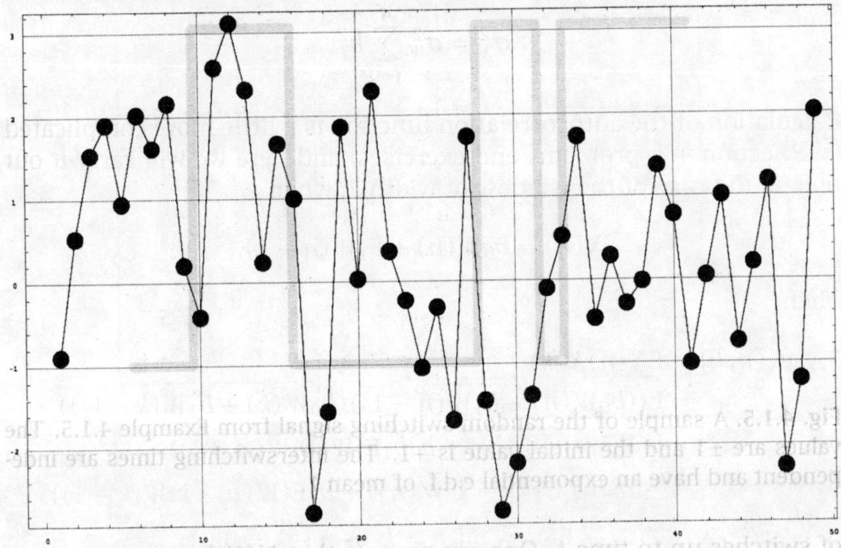

Fig. 4.1.3. Sample moving average signal $X(n) = 2W(n) + 5W(n-1)$ for the sample white noise shown in Figure 4.1.2. Note that the moving average signal appears smoother than the original white noise. The constrained oscillations are a result of nontrivial, although short-term in this example, correlations.

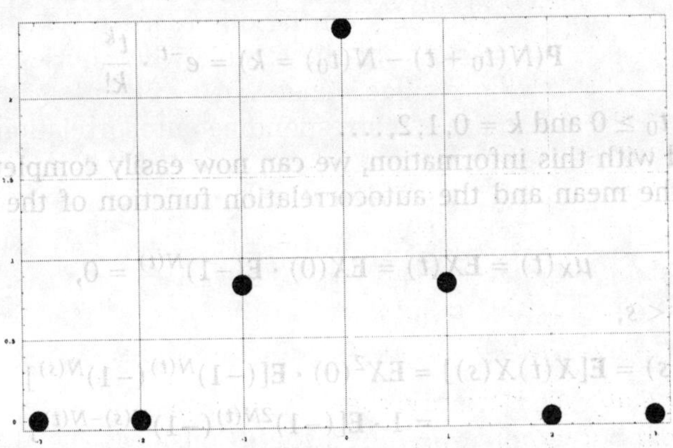

Fig. 4.1.4. Autocovariance function for the moving average signal $X(n) = 2W(n) + 5W(n-1)$. Note that the values of the signal separated by more that one time unit are uncorrelated.

Calculation of the mean and the autocorrelation function of the switching signal depends on the knowledge of the fact that such a random signal can be written in the form

$$X(0) \cdot (-1)^{N(t)},$$

where $N(t)$ is the (nonstationary) random signal counting the number

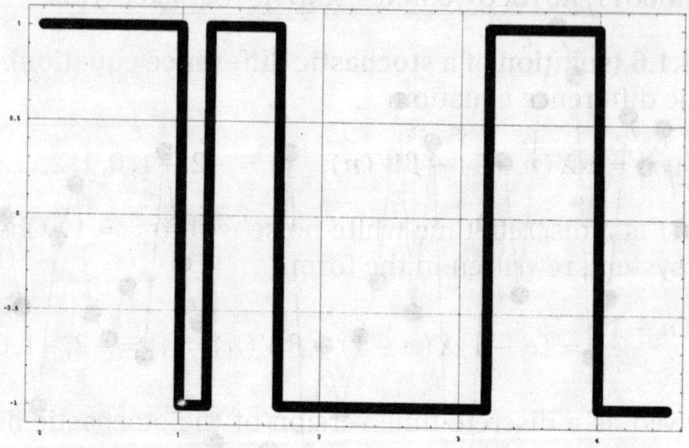

Fig. 4.1.5. A sample of the random switching signal from Example 4.1.5. The values are ± 1 and the initial value is $+1$. The interswitching times are independent and have an exponential c.d.f. of mean 1.

of switches up to time t. One can prove[27] that $N(t)$ has increments in disjoint time intervals that are statistically independent, with the distributions thereof depending only on the interval's length. More strikingly, these increments must have the Poisson probability distribution with mean equal to the interval's length, that is,

$$\mathbf{P}(N(t_0+t) - N(t_0) = k) = e^{-t} \cdot \frac{t^k}{k!}$$

for any $t, t_0 \geq 0$ and $k = 0, 1, 2, \ldots$.

Armed with this information, we can now easily complete calculations of the mean and the autocorrelation function of the switching signal:

$$\mu_X(t) = \mathbf{E}X(t) = \mathbf{E}X(0) \cdot \mathbf{E}(-1)^{N(t)} = 0,$$

and, for $t < s$,

$$\begin{aligned}\gamma_X(t,s) = \mathbf{E}[X(t)X(s)] &= \mathbf{E}X^2(0) \cdot \mathbf{E}[(-1)^{N(t)}(-1)^{N(s)}] \\ &= 1 \cdot \mathbf{E}[(-1)^{2N(t)}(-1)^{N(s)-N(t)}] \\ &= \mathbf{E}(-1)^{N(s)-N(t)} \\ &= \sum_{k=0}^{\infty}(-1)^k \cdot \frac{e^{-(s-t)}(s-t)^k}{k!} = e^{-2(s-t)}.\end{aligned}$$

Therefore, the random switching signal $X(t)$ is stationary and, because of the symmetry property of all autocorrelation functions, its autocorrelation function

$$\gamma_X(t) = e^{-2|t|}.$$

---

[27] See, for example, O. Kallenberg, *Foundations of Modern Probability*, Springer-Verlag, New York, 1997.

Example 4.1.6 (solution of a stochastic difference equation). Consider a stochastic difference equation

$$X(n) = \alpha X(n-1) + \beta W(n), \quad n = -2, -1, 0, 1, 2, \ldots,$$

where $W(n)$ is a discrete-time white noise with $\sigma_W^2 = 1$. Observe that the above system, rewritten in the form

$$\frac{X(n) - X(n-1)}{\Delta n} = (\alpha - 1)X(n-1) + \beta W(n), \quad n = -2, -1, 0, 1, 2, \ldots,$$

can be viewed as a discrete-time version of the stochastic differential equation

$$dX(t) = (\alpha - 1)X(t)dt + \beta W(t)dt,$$

where $W(t)$ represents the continuous-time version of the white noise to be discussed in later chapters.

The solution of the above stochastic difference equation can be found by recursion. Therefore,

$$\begin{aligned} X(n) &= \alpha(\alpha X(n-2) + \beta W(n-1)) + \beta W(n) \\ &= \alpha^2 X(n-2) + \alpha\beta W(n-1) + \beta W(n) \\ &= \cdots = \alpha^l X(n-l) + \sum_{k=0}^{l-1} \alpha^k \beta W(n-k). \end{aligned}$$

for any $l = 1, 2, \ldots$. Assuming that $|\alpha| < 1$ and that $X(n-k)$ remain bounded, the first term $\alpha^k X(n-k) \to 0$ as $k \to \infty$. In that case, the second term converges to the infinite sum and the solution is of the form

$$X(n) = \beta \sum_{k=0}^{\infty} \alpha^k W(n-k).$$

This is the special form of the general moving average signal appearing in Problem 4.3.4, with the windowing sequence

$$c_k = \begin{cases} \beta\alpha^k & \text{for } k = 0, 1, 2, \ldots; \\ 0 & \text{for } k = -1, -2, \ldots. \end{cases}$$

Hence its autocorrelation function is

$$\gamma_X(n) = \sum_{k=-\infty}^{\infty} c_k c_{n+k} = \beta^2 \sum_{k=0}^{\infty} \alpha^k \alpha^{n+k} = \beta^2 \frac{\alpha^n}{1 - \alpha^2}.$$

Example 4.1.7 (using moving averages to filter noise out of a signal). Consider a signal of the form

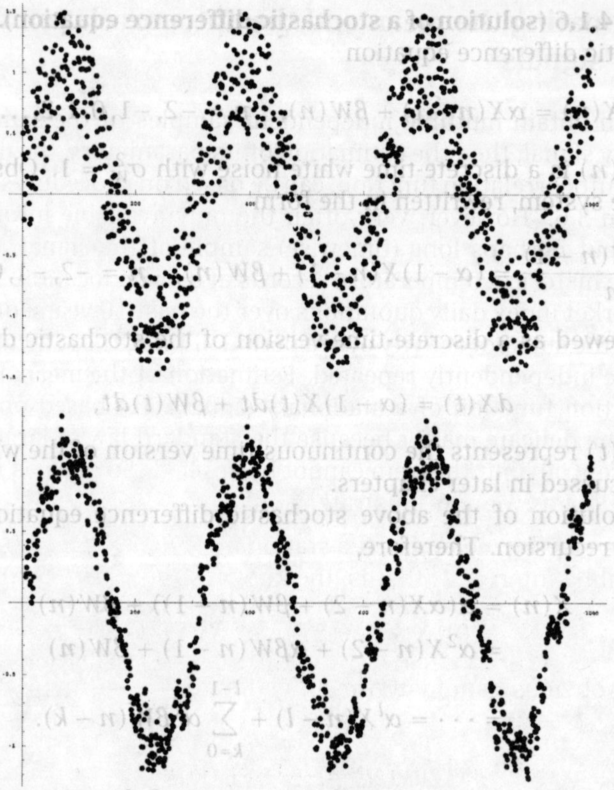

Fig. 4.1.6. *Top*: Signal $X(n)$ from Example 4.1.7 containing a nonrandom harmonic component plus a random white noise. *Bottom*: The same signal after a smoothing, moving average operation filtered out some of the white noise. The figure shows values of both signals for times $n = 1, 2, \ldots, 1000$.

$$X(n) = \sin(0.02n) + W(n),$$

where $W(n)$ is the white noise considered in Example 4.1.3 (shown in Figure 4.1.2), and let $Y(n)$ be a moving average (discrete-time convolution) of signal $X(n)$ with the windowing sequence $b_0 = b_1 = b_2 = b_3 = b_4 = \frac{1}{5}$, that is,

$$Y(n) = \frac{1}{5}X(n) + \frac{1}{5}X(n-1) + \frac{1}{5}X(n-2) + \frac{1}{5}X(n-3) + \frac{1}{5}X(n-4).$$

The values of both signals, $X(n)$ and $Y(n)$, for time instants $n = 1, 2, \ldots, 1000$, are shown in Figure 4.1.6. Clearly, the moving average operation filtered some of the white noise out of the original signal and the transformed signal appears smoother.

4.2 Estimating the mean and the autocorrelation function, ergodic signals

If one can obtain multiple independent samples of the same random stationary signal, then the estimation of its parameters, the mean value and the autocorrelation function, can be based on procedures described in Section 3.6. However, very often the only available information is a single but perhaps long (timewise) sample of the signal; think here about the historical temperature records at a given location, Dow Jones stock market index daily quotations over the past 10 years, or measurements of the sunspot activity over a period of time; these measurements cannot be independently repeated. Estimation of the mean and the autocorrelation function of a stationary signal $X(t)$ based on its single sample is a delicate matter because the standard law of large numbers and the central limit theorem cannot be applied. So one has to proceed with caution, as we now illustrate.

Estimation of the mean μ_X. If a stationary signal $X(t)$ is sampled with the sampling interval T, that is, the known values are

$$X(0), X(T), X(2T), \ldots, X(NT), \ldots,$$

then the obvious candidate for an estimator $\hat{\mu}_X$ of the signal's mean μ_X is

$$\hat{\mu}_X(N) = \frac{1}{N}\sum_{i=0}^{N-1} X(iT).$$

This estimator is easily seen to be unbiased as

$$\mathbf{E}[\hat{\mu}_X(N)] = \frac{1}{N}\sum_{i=0}^{N-1} \mathbf{E}[X(iT)] = \mu_X. \tag{4.2.1}$$

To check whether the estimator $\hat{\mu}_X(N)$ converges to μ_X as the observation interval $NT \to \infty$, that is, to check the estimator's consistency, we will take a look at the mean-square distance (estimation error) between $\hat{\mu}_X(N)$ and μ_X or, equivalently, the variance of their difference:

$$\begin{aligned}
\sigma^2(\hat{\mu}_X(N)) &= \mathbf{E}[(\hat{\mu}_X - \mu_X)^2] \\
&= \frac{1}{N^2}\mathbf{E}\left[\sum_{i=0}^{N-1}(X(iT) - \mu_X)\sum_{k=0}^{N-1}(X(kT) - \mu_X)\right] \\
&= \frac{1}{N^2}\sum_{i=0}^{N-1}\sum_{k=0}^{N-1}\gamma_X(iT, kT) = \frac{1}{N^2}\sum_{i=0}^{N-1}\sum_{k=0}^{N-1}\gamma_C((i-k)T) \\
&= \frac{\sigma_X^2}{N} + \frac{2}{N}\sum_{k=0}^{N-1}\left(1 - \frac{k}{N}\right)\gamma_X(kT).
\end{aligned} \tag{4.2.2}$$

So the error of replacing the true value μ_X by the estimator $\hat{\mu}_X$ will converge to zero, as $N \to \infty$, only if the sum in (4.2.2) increases more slowly[28] than N, i.e.,

$$\sum_{k=0}^{N-1}\left(1-\frac{k}{N}\right)\gamma_X(kT) = o(N) \quad \text{as } N \to \infty. \tag{4.2.3}$$

Thus, for example, if the covariance function $\gamma_X(n)$ vanishes outside a finite interval, as was the case for finite moving averages in Example 4.1.2, then $\hat{\mu}_X$ is a consistent estimator for μ_X.

Example 4.2.1 (consistency of $\hat{\mu}_X$ for solutions of discrete-time stochastic difference equations). Consider the solution $X(n)$ of the stochastic difference equation from Example 4.1.6. Its autocorrelation function was found to be of the form

$$\gamma_X(n) = \beta^2\frac{\alpha^n}{1-\alpha^2}, \quad |\alpha| < 1.$$

Since it decays exponentially as $n \to \infty$, the sum in (4.2.2) converges and condition (4.2.3) is satisfied. The mean-square error of replacing μ_X by the estimator $\hat{\mu}_X$ can now be controlled:

$$\begin{aligned}\sigma^2(\hat{\mu}_X(N)) &= \mathbf{E}[(\hat{\mu}_X - \mu_X)^2]\\ &= \frac{\gamma_X(0)}{N} + \frac{2}{N}\sum_{k=0}^{N-1}\left(1-\frac{k}{N}\right)\beta^2\frac{\alpha^k}{1-\alpha^2}\\ &\le \frac{\beta^2}{N(1-\alpha^2)}\left(1+2\sum_{k=0}^{N-1}\alpha^k\right) \le \frac{\beta^2(3-\alpha)}{N(1-\alpha^2)(1-\alpha)}.\end{aligned}$$

Estimation of the covariance function $\gamma_X(n)$. For simplicity's sake assume that $\mu_X = 0$, the sampling interval $T = 1$, the signal is real-valued, and that observations $X(0), \ldots, X(N)$ are given. The natural candidate for an estimator of the autocorrelation function $\gamma_X(n) = \mathbf{E}X(0)X(n)$ is the time average:

$$\hat{\gamma}_X(n;N) = \frac{1}{N-n}\sum_{k=0}^{N-n-1} X(k)X(k+n). \tag{4.2.4}$$

It is an unbiased estimator since

$$\mathbf{E}[\hat{\gamma}_X(n,N)] = \frac{1}{N-n}\mathbf{E}\left[\sum_{k=0}^{N-n-1} X(k)X(k+n)\right]$$

[28] Here we use Landau's asymptotic notation: we write that $f(x) = o(g(x))$, as $x \to x_0$, and say that $f(x)$ is little "oh" of $g(x)$ at x_0, if $\lim_{x\to x_0}\frac{f(x)}{g(x)} = 0$.

$$= \frac{1}{N-n} \sum_{k=0}^{N-n-1} \gamma_X(n) = \gamma_X(n).$$

One can also prove that if $\gamma_X(n) \to 0$ sufficiently fast,[29] as $n \to \infty$, and if $\gamma_X(0) = \sigma_X^2 < \infty$, then the mean-square distance from $\hat{\gamma}_X(n;N)$ to $\gamma_X(n)$ decreases to 0 as $N \to \infty$. In other words, the estimator (4.2.4) is consistent.

Example 4.2.2. Figure 4.2.1 shows two samples of the central channel recording for a full-term neonate EEG (see Figure 4.1.1 for a sample of the full 21-channel EEG). The duration of each of the samples is three minutes. The data in the top picture were recorded during the active sleep stage, and in the bottom picture during the quiet sleep stage. The estimated autocorrelation functions (ACFs) for both signals were then calculated using formula (4.2.4), and are shown in Figure 4.2.2. The example is taken from A. Piryatinska's Ph.D. dissertation (Department of Statistics, Case Western Reserve University, Cleveland, 2004), mentioned already in Section 4.1. Note that the ACF of the active sleep signal decays much more slowly than the ACF of the quiet sleep, indicating the longer-range dependence structure of the former. Information on the rate of decay in EEG ACFs can then be used to automatically classify stationary segments of the EEG signals as those corresponding to different sleep stages recognized by pediatric neurologists.

Remark 4.2.1 (*ergodicity*). If the estimator $\hat{\mu}_X$ is unbiased and consistent, that is,

$$\mathbf{E}\hat{\mu}_X(N) = \mu_X \quad \text{and} \quad \sigma^2(\hat{\mu}_X(N)) \to 0,$$

as $N \to \infty$, then one often says that the signal is *ergodic in the mean*. Note that, in general, this does not imply that for every sample path of the random signal the estimator converges to the estimated parameter. To guarantee that, for a general test function g, the time averages

$$\frac{g(X(1)) + g(X(2)) + \cdots + g(X(N))}{N}$$

converge to $\mathbf{E}g(X(1))$ as $N \to \infty$, for (almost) every sample path of the random signal, stronger ergodicity and stricter stationarity assumptions are needed. A more detailed analysis of the ergodic behavior of stationary time series can be found in the above-quoted books by M. Denker and W. A. Woyczyński and by P. J. Brockwell and R. A. Davis.

[29] For a thorough exposition of these issues, see, for example, P. J. Brockwell and R. A. Davis, *Time Series: Theory and Methods*, Springer-Verlag, New York, 1991.

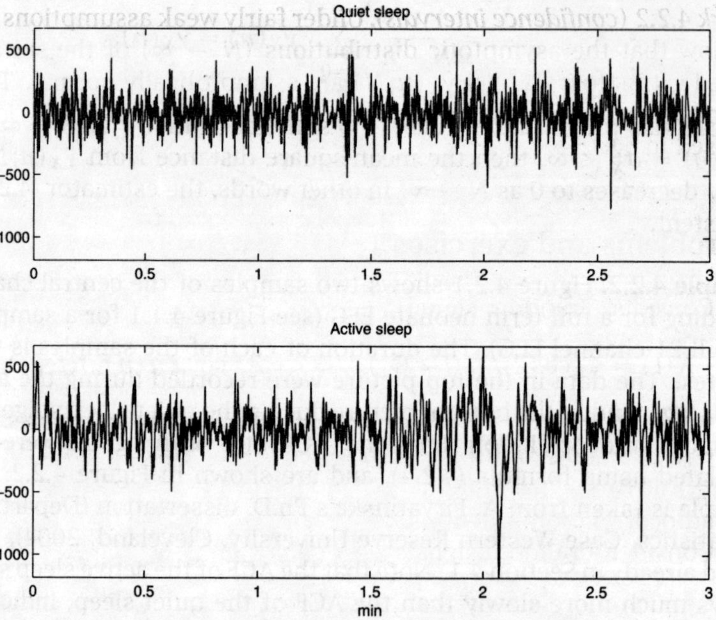

Fig. 4.2.1. *Top*: Three-minute recording of the central channel EEG for an infant in a quiet sleep stage. *Bottom*: Analogous recording for an active sleep stage.

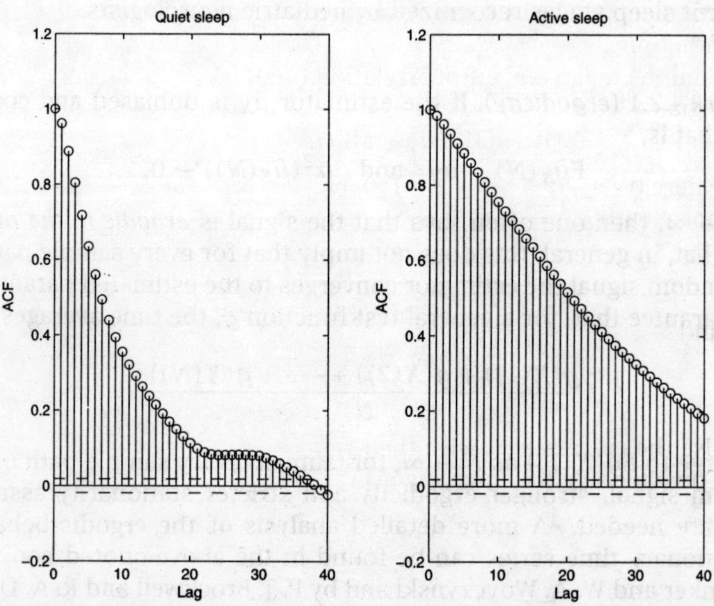

Fig. 4.2.2. *Left*: Estimated autocorrelation function (ACF) for the quiet sleep EEG signal from Figure 4.2.1. *Right*: Analogous estimated ACF for the active sleep stage.

Remark 4.2.2 (*confidence intervals*). Under fairly weak assumptions one can show that the asymptotic distributions ($N \to \infty$) of the suitably rescaled estimators $\hat{\mu}_X(N)$, $\hat{\gamma}_X(n;N)$ are asymptotically normal. Thus the confidence intervals for them can be constructed following the ideas discussed in Section 3.6.

4.3 Problems and exercises

4.3.1. Consider a random signal

$$X(t) = A_1 \cos 2\pi f_0(t + \Theta_1) + \cdots + A_n \cos 2\pi(nf_0)(t + \Theta_n),$$

where $A_1, \Theta_1, \ldots, A_n, \Theta_n$ are independent random variables and $\Theta_1, \ldots, \Theta_n$ are uniformly distributed on the time interval $[0, P = \frac{1}{f_0}]$. Is this signal stationary? Find its mean, autocovariance, and autocorrelation functions.

4.3.2. Consider a random signal

$$X(t) = A_1 \cos 2\pi f_0(t + \Theta_0),$$

where A_1, Θ_0, are independent random variables, and Θ_0 is uniformly distributed on the time interval $[0, \frac{P}{3} = \frac{1}{3f_0}]$. Is this signal stationary? Find its mean, autocovariance, and autocorrelation functions.

4.3.3. Find the mean and autocorrelation functions of the discrete-time signal

$$Y(n) = 3W(n) + 2W(n-1) - W(n-2),$$

where $W(n), n = \ldots, -2, -1, 0, 1, 2, \ldots$, is the discrete-time white noise with $\sigma_W^2 = 4$, that is,

$$EW(n) = 0$$

and

$$\mathrm{E}(W(k)W(n)) = 4\delta(n-k) = \begin{cases} 4 & \text{if } n-k=0; \\ 0 & \text{if } n-k \neq 0. \end{cases}$$

Use the calculations with the Kronecker δ explicitly.

4.3.4. Consider a general moving average signal

$$X(n) = \sum_{k=-\infty}^{\infty} c_k W_{n-k},$$

where c_k is a "windowing" sequence such that $\sum_k |c_k|^2 < \infty$, and $W(n)$ is the standard white noise signal with $\gamma_W(n) = \delta(n)$. Show that the covariance function is

$$\gamma_X(n) = \sum_{k=-\infty}^{\infty} c_k c_{n+k}.$$

Use the calculations with the Kronecker δ explicitly. Apply this formula to verify the solution to Problem 4.3.3.

4.3.5. *Simulation of white noise with an arbitrary probability distribution.* Formula (3.1.11), $F_Y(y) = F_X(g^{-1}(y))$, describes the c.d.f. $F_Y(y)$ of the random quantity $Y = g(X)$ in terms of the c.d.f. $F_X(x)$ of the random quantity X and the function $g(x)$. It also permits construction of an algorithm to produce random samples from any given probability distribution provided a random sample uniformly distributed on the interval $[0, 1]$ is given. The latter can be obtained by using the random number generator in any computing platform; see Problem 1.4.15.

Let U be a uniformly distributed on $[0, 1]$ random quantity U with the c.d.f.

$$F_U(u) = u, \quad 0 \le u \le 1, \tag{4.3.1}$$

Then for a given c.d.f. $F_Z(z)$, the random quantity $Z = F_Z^{-1}(U)$, where $F_Z^{-1}(u)$ is the function inverse to $F_Z(z)$ (that is, a solution of the equation $u = F_Z(F_Z^{-1}(u))$), has the c.d.f. $F_Z(z)$. Indeed, a simple calculation using (4.3.1) shows that

$$\mathbf{P}(F_Z^{-1}(U) \le z) = \mathbf{P}(U \le F_Z(z)) = F_Z(z)$$

because $0 \le F_Z(z) \le 1$. So, for example, if the desired c.d.f. is exponential, with $F_Z(z) = 1 - e^{-z}, z \ge 0$, then $F_Z^{-1}(u) = -\ln(1 - u), 0 \le u \le 1$, and the random quantity $Z = -\ln(1 - U)$ has the above exponential c.d.f.

The general simulation algorithm is thus as follows:

(i) Choose the sample size N, and produce a random sample, $u_1, u_2, \dots, u_N$, uniformly distributed on $[0, 1]$.

(ii) Calculate the inverse function $F_Z^{-1}(u)$.

(iii) Substitute the random sample, $u_1, u_2, \dots, u_N$, into $F_Z^{-1}(u)$ to obtain the random sample

$$z_1 = F_Z^{-1}(u_1), z_2 = F_Z^{-1}(u_2), \dots, z_N = F_Z^{-1}(u_N),$$

which has the desired c.d.f. $F_Z(z)$.

Use the above algorithm and Problem 1.4.15 to produce and plot examples of the white noise $W(n)$ with (a) the double exponential p.d.f. $f_W(w) = \frac{e^{-|w|}}{2}$ and (b) the Cauchy p.d.f. $f_W(w) = (\pi(1 + w^2))^{-1}$. Start with a calculation of the corresponding c.d.f.s. Check the result graphically by plotting the histograms of the random samples against the theoretical p.d.f.s.

4.3.6. Simulations of stationary random signals. Using the algorithm from Problem 4.3.5, repeat simulations shown in Figures 4.1.2, 4.1.3, and 4.1.6, but replacing the uniformly distributed white noise by (a) a double exponentially distributed white noise and (b) a "white noise" with the Cauchy distribution. Experiment with these simulations by including parameters in the above p.d.f.s, and changing the length of the produced discrete-time random signals.

4.3.7. Using the procedures described in Section 4.2, estimate the means and the autocorrelation functions (ACF) for sample signals obtained in simulation in Problem 4.3.6(a). Then compare graphically the estimated and the theoretical ACFs.

Note. Cauchy random quantities have an infinite variance (check!—cf. Problem 3.7.20), so the correlational definition of the discrete-time white noise is not applicable for them. In such cases, by a discrete-time white noise $W(n), \ldots, -2, -1, 0, 1, 2, \ldots$, we simply mean a sequence of independent, identically and symmetrically distributed (i.e., $W(n) \sim -W(n)$) random quantities. No moment requirements are made. On the other hand, such a sequence always forms a strictly stationary random signal; cf. (3.3.24) and Problem 3.7.28.

5

Power Spectra of Stationary Signals

The Fourier transform $X(f)$ of the sample paths of a stationary random signal $X(t)$ does not exist in the usual sense, and analysis of the spectral contents of such signals requires a different, more subtle approach which has to rely on the concept of the *mean power* of the random signal. Only then we can investigate how the energy is distributed over different frequencies. The question is, of course, of fundamental importance in practical applications, as real-life signal processing devices such as measuring instruments, amplifiers, antennas, etc., transmit different frequencies with different attenuation.

5.1 Mean power of a stationary signal

Mean energy $\mathbf{E}(E_X)$ of a stationary signal $X(t)$, that is, the expected value of the energy, is infinite. Indeed, using the linearity property of expectations we can interchange the order of taking the mean and the integration to obtain that

$$\mathbf{E}[E_X] = \mathbf{E}\int_{-\infty}^{\infty} X^2(t)\,dt = \int_{-\infty}^{\infty} \mathbf{E}(X^2(t))\,dt = \int_{-\infty}^{\infty} \sigma_X^2\,dt = \infty. \tag{5.1.1}$$

However, the mean power $\mathbf{E}(\mathrm{PW}_X)$ of a stationary signal is always finite since

$$\mathbf{E}[\mathrm{PW}_X] = \mathbf{E}\lim_{T\to\infty}\frac{1}{2T}\int_{-T}^{T} X^2(t)\,dt = \sigma_X^2 < \infty. \tag{5.1.2}$$

To find the distribution of power PW_X over different frequencies f, we will consider a windowed signal

$$X_T(t) = \begin{cases} X(t) & \text{for } |t| \le T; \\ 0 & \text{otherwise,} \end{cases} \tag{5.1.3}$$

that is the original signal restricted to the time window of duration $2T$. Then, in view of the Parseval equality established in Section 2.4, with the Fourier transform of the windowed signal defined by the equality

$$X_T(f) = \int_{-\infty}^{\infty} X_T(t)e^{-j2\pi ft}dt = \int_{-T}^{T} X(t)e^{-j2\pi ft}dt,$$

we can express the mean power of the original signal by the formula

$$\begin{aligned}
\mathbf{E}[\mathrm{PW}_X] &= \mathbf{E}\left[\lim_{T\to\infty}\frac{1}{2T}\int_{-T}^{T} X^2(t)dt\right] \\
&= \mathbf{E}\left[\lim_{T\to\infty}\frac{1}{2T}\int_{-\infty}^{\infty} X_T^2(t)dt\right] = \mathbf{E}\left[\lim_{T\to\infty}\frac{1}{2T}\int_{-\infty}^{\infty} |X_T(f)|^2 df\right] \\
&= \int_{-\infty}^{\infty}\lim_{T\to\infty}\frac{\mathbf{E}|X_T(f)|^2}{2T}df.
\end{aligned}$$

Denoting

$$S_X(f) = \lim_{T\to\infty}\frac{\mathbf{E}|X_T(f)|^2}{2T}, \tag{5.1.4}$$

the mean power has the representation

$$\mathbf{E}(\mathrm{PW}_X) = \int_{-\infty}^{\infty} S_X(f)df. \tag{5.1.5}$$

The function $S_X(f)$ is called the *power spectral density* or, simply, the *power spectrum* of the stationary signal $X(t)$. It shows how the mean power PW_X of the random stationary signal $X(t)$ is distributed over different frequencies $f, -\infty < f < \infty$. The mean power concentrated in a frequency band $f_1 < f < f_2$ is then given by the integral

$$\int_{f_1}^{f_2} S_X(f)df.$$

5.2 Power spectrum and autocorrelation function

What makes the power spectrum $S_X(f)$ a practical tool in the analysis of random stationary signals is the fact that it is simply the Fourier transform of the signal's autocorrelation function $\gamma_X(t)$. In other words,

$$S_X(f) = \int_{-\infty}^{\infty} \gamma_X(t)e^{-j2\pi ft}dt. \tag{5.2.1}$$

This fundamental property can be easily verified by direct calculation. Indeed,

$$\begin{aligned}
S(f) &= \lim_{T\to\infty} \frac{\mathbf{E}|X_T(f)|^2}{2T} = \lim_{T\to\infty} \frac{\mathbf{E}(X_T(f)X_T^*(f))}{2T} \\
&= \lim_{T\to\infty} \frac{1}{2T}\mathbf{E}\left[\int_{-T}^{T} X(t)e^{-2\pi jft}dt \int_{-T}^{T} X(s)e^{2\pi jfs}ds\right] \\
&= \lim_{T\to\infty} \frac{1}{2T}\int_{-T}^{T}\int_{-T}^{T} \mathbf{E}[X(t)X(s)]e^{-2\pi jf(t-s)}dtds \\
&= \lim_{T\to\infty} \frac{1}{2T}\left[\int_{-T}^{T}\left(\int_{-T-s}^{T-s} \gamma_X(u)e^{-2\pi jfu}du\right)ds\right] \\
&= \int_{-\infty}^{\infty} \gamma_X(u)e^{-2\pi jfu}du.
\end{aligned}$$

Given the properties of the Fourier transform and its inverse, we also have the fact that the autocorrelation $\gamma_X(\tau)$ of signal $X(t)$ is the inverse Fourier transform of the power spectrum $S_X(f)$:

$$\gamma_X(t) = \int_{-\infty}^{\infty} S_X(f)e^{j2\pi ft}df.$$

Estimation of the power spectrum $S_X(f)$. For simplicity's sake assume that the signal is real-valued, and that observations $X(0),\ldots,X(N)$ are given. To estimate the spectrum the natural way to proceed is to replace the theoretical ACF $\gamma_X(t)$ in (5.2.1) by the estimated ACF $\hat{\gamma}_X(n;N)$ given by the formula (4.2.4) and replace the integral by the finite sum. This yields the estimator:

$$\begin{aligned}
\hat{S}_X(f;N) &= \sum_{n=-(N-1)}^{N-1} \hat{\gamma}_X(|n|;N)e^{-j2\pi fn} \\
&= \sum_{n=-(N-1)}^{N-1} \frac{1}{N-|n|}\sum_{k=0}^{N-|n|-1} X(k)X(k+|n|)e^{-j2\pi fn},
\end{aligned}$$

which then, for large N, can be rewritten (see Section 9.2) in the form

$$\hat{S}_X(f;N) \approx I_N(f) := \frac{1}{N}\left|\sum_{n=1}^{N} X(n)e^{-j2\pi fn}\right|^2. \tag{5.2.2}$$

The quantity $I_N(f)$ is usually called the *periodogram* of the sampled signal $X(t)$ based on a sample of size N.

Let us return now to two samples of neonatal sleep signals displayed in Figure 4.2.1. Their estimated autocorrelation functions were shown in Figure 4.2.2. Their periodograms have been calculated using formula (5.2.2) and are reproduced in Figure 5.2.1. Since the signal was recorded at the sampling frequency of 64 Hz, and the duration of each recording was three minutes, the total number of sample points is $N = 192$. The reader will notice that the periodogram is quite noisy and, perhaps, should be smoothed out to better reflect the true spectrum of the random signal. Nevertheless, a comparison of these rough spectra for quiet sleep and active sleep segments clearly shows that the active sleep signal shows greater concentration of the spectrum at low frequencies than the quite sleep signal.

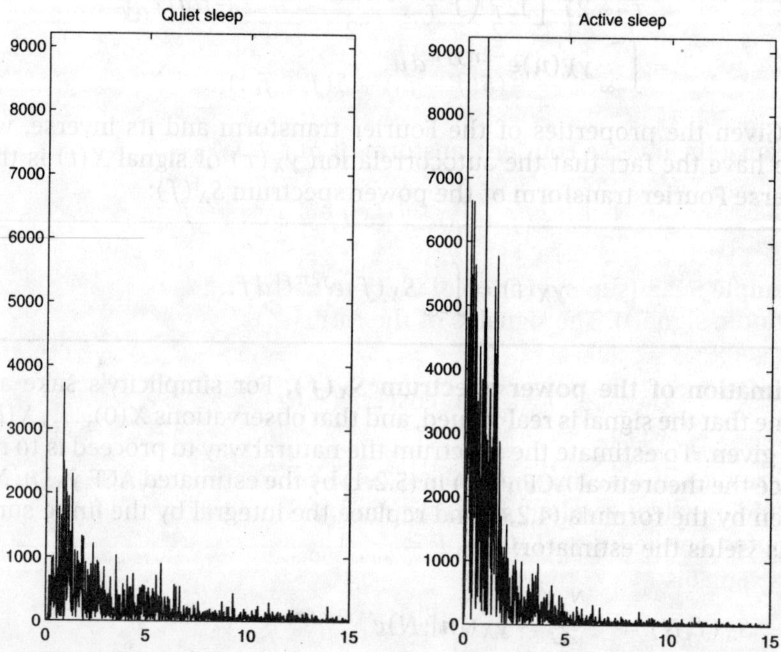

Fig. 5.2.1. *Left*: Periodogram of the neonatal quiet sleep EEG signal from Figure 4.2.1. *Right*: Analogous periodogram for the active sleep stage. (From A. Piryatinska's Ph.D. dissertation, Department of Statistics, Case Western Reserve University, Cleveland, 2004.)

Example 5.2.1 (simple random harmonic oscillation). In this case, the random signal is of the form

$$X(t) = A\cos(2\pi f_0(t + \Theta)),$$

where the random amplitude A has zero mean, $EA = 0$, and finite variance $EA^2 < \infty$. The random phase Θ is independent of A and uniformly distributed on the interval $[0, P]$ with $P = \frac{1}{f_0}$. In Chapter 4, we calculated that the autocorrelation function for this signal is

$$\gamma_X(\tau) = \frac{\mathbf{E}|A|^2}{2}\cos(2\pi f_0 \tau).$$

Hence the power spectrum of the simple random harmonic oscillation with fundamental frequency f_0 is

$$\begin{aligned} S_X(f) &= \int_{-\infty}^{\infty} \gamma_X(\tau) e^{-2\pi j f \tau}\, d\tau \\ &= \int_{-\infty}^{\infty} \frac{\mathbf{E}|A|^2}{2} \frac{e^{j2\pi f_0 \tau} + e^{-2\pi j f_0 \tau}}{2} e^{-j2\pi f \tau}\, d\tau \\ &= \frac{\mathbf{E}|A|^2}{4}(\delta(f - f_0) + \delta(f + f_0)), \end{aligned}$$

because the inverse Fourier transform of $\delta(f - f_0)$ is

$$\int_{-\infty}^{\infty} \delta(f - f_0) e^{2\pi j f \tau}\, df = e^{2\pi j f_0 \tau}.$$

Example 5.2.2 (superposition of random harmonic oscillations (random periodic signal)). The signal is of the form

$$X(t) = \sum_{k=1}^{N} A_k \cos(2\pi k f_0 (t + \Theta_k)),$$

where the zero-mean amplitudes $A_1, \ldots, A_N$ and phases $\Theta_1, \ldots, \Theta_N$ are all independent random quantities and $\Theta_1, \ldots, \Theta_N$ are uniformly distributed on the interval $[0, P]$, $P = \frac{1}{f_0}$. The autocovariance function of this signal is

$$\gamma_X(\tau) = \sum_{k=1}^{N} \frac{\mathbf{E}|A|^2}{2} \cos(2\pi f_0 k \tau),$$

and, arguing as in Example 5.2.1, the power spectrum is a linear combination of the Dirac deltas:

$$S_X(f) = \frac{1}{4} \sum_{k=1}^{N} \mathbf{E}|A_k|^2 (\delta(f - k f_0) + \delta(f + k f_0))$$

Thus in this case the power spectrum is concentrated on discrete frequencies $f_0, 2f_0, \ldots, N f_0$.

Example 5.2.3 (the continuous-time white noise signal). By the standard white noise signal, we mean a signal $W(t)$ with a totally flat power spectrum

$$S_W(f) = 1, \quad -\infty < f < \infty.$$

Clearly, such a signal is not realizable physically since its mean power is infinite:

$$\mathrm{E}(\mathrm{PW}_W) = \int_{-\infty}^{\infty} 1 df = \infty.$$

However, it is a very useful abstraction. The Fourier transform of its autocorrelation function $\gamma_W(\tau)$ must satisfy the equation

$$\int_{-\infty}^{\infty} \gamma_W(\tau) e^{-j2\pi f\tau} d\tau \equiv 1$$

for all $-\infty < f < \infty$, which implies that

$$\gamma_W(\tau) = \delta(\tau).$$

Loosely speaking, the above formula can be interpreted as follows: we can say that, for $t \neq s$, the white noise has values $X(t)$ and $X(s)$ that are uncorrelated and, for $t = s$, the correlation between $X(t)$ and $X(s)$ is infinite. This autocorrelation function is thus not a true function, but its Dirac delta impulse shape is not surprising if you compare it to the shape of the autocorrelation function for the discrete-time white noise discussed in Chapter 4. Because of the form of its autocorrelation function, the white noise is sometimes called a *delta-correlated signal.*

If a random signal $W(t)$ has the spectrum $S_W(f) \equiv \mathcal{N}_0 > 0$, then we shall call $W(t)$ a white noise of amplitude $\mathcal{N}_0$.

Example 5.2.4 (band-limited noise). The white noise model can be adjusted to produce a physically realizable signal by assuming that the power spectrum is flat but only over a limited band of spectrum frequencies. More precisely, we shall call a signal $X(t)$ a *band-limited white noise* if its spectrum is

$$S_X(f) = \begin{cases} \mathcal{N}_0 & \text{for } -f_{\max} < f < f_{\max}; \\ 0 & \text{elsewhere.} \end{cases}$$

The mean power of the band-limited white noise is finite:

$$\mathrm{PW}_X = \int_{-\infty}^{\infty} S_X(f) df = \int_{-f_{\max}}^{f_{\max}} \mathcal{N}_0 df = 2 f_{\max} \mathcal{N}_0.$$

The autocorrelation function of the band-limited white noise can be easily calculated by taking the inverse Fourier transform. Thus we obtain

$$\begin{aligned} \gamma_X(\tau) &= \int_{-\infty}^{\infty} S_X(f) e^{j2\pi f\tau} df = \mathcal{N}_0 \int_{-f_{\max}}^{f_{\max}} e^{j2\pi f\tau} df \\ &= \frac{\mathcal{N}_0}{j2\pi\tau}\left(e^{j2\pi f_{\max}\tau} - e^{-j2\pi f_{\max}\tau}\right) = \frac{\mathcal{N}_0}{\pi\tau} \sin(2\pi f_{\max}\tau). \end{aligned}$$

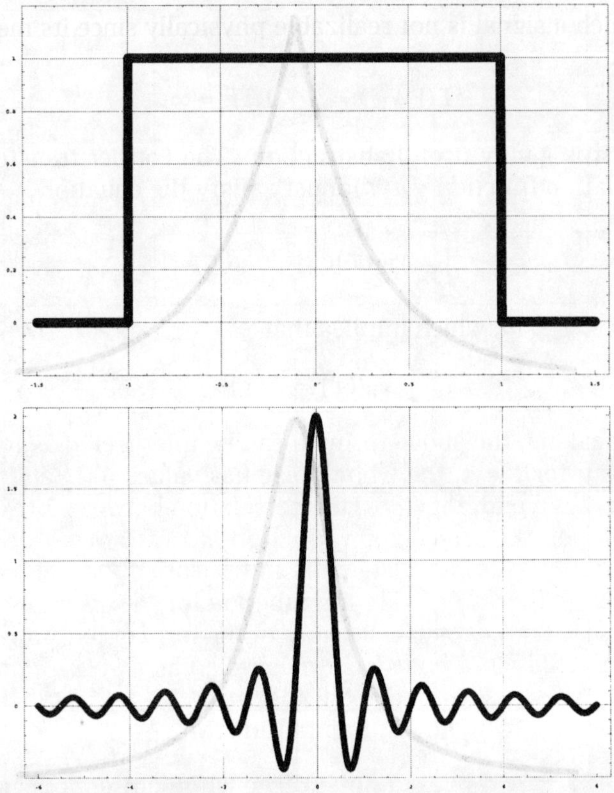

Fig. 5.2.2. *Top*: Power spectrum of the band-limited white noise $X(t)$ from Example 5.2.4. The bandwidth is $2f_{\max}$ and the mean power is $\mathrm{PW}_X = 2$. *Bottom*: Autocorrelation function of the above band-limited noise. Observe that as the bandwidth expands to infinity the autocorrelation function approaches the Dirac delta, the autocorrelation function of ideal white noise.

Figure 5.2.2 shows both the power spectrum of a band-limited white noise and its autocorrelation function for $f_{\max} = 1$ and $\mathcal{N}_0 = 1$. Observe that, not surprisingly, as the bandwidth $2f_{\max}$ expands to infinity the autocorrelation function approaches the Dirac delta, the autocorrelation function of the ideal white noise. Note that the maximum value of the autocorrelation function $\gamma_X(\tau)$ is attained at $\tau = 0$ and is equal to the mean power $\mathrm{PW}_X = 2f_{\max}$ which diverges to $+\infty$ as the bandwidth increases. However, $\int_{-\infty}^{\infty} \gamma_X(\tau)d\tau = S_X(0) = 1$, and the value of the power spectrum at zero frequency is independent of the bandwidth and remains constant.

Example 5.2.5 (random switching signal). The random switching signal $X(t)$ discussed in Chapter 4 has the autocorrelation function

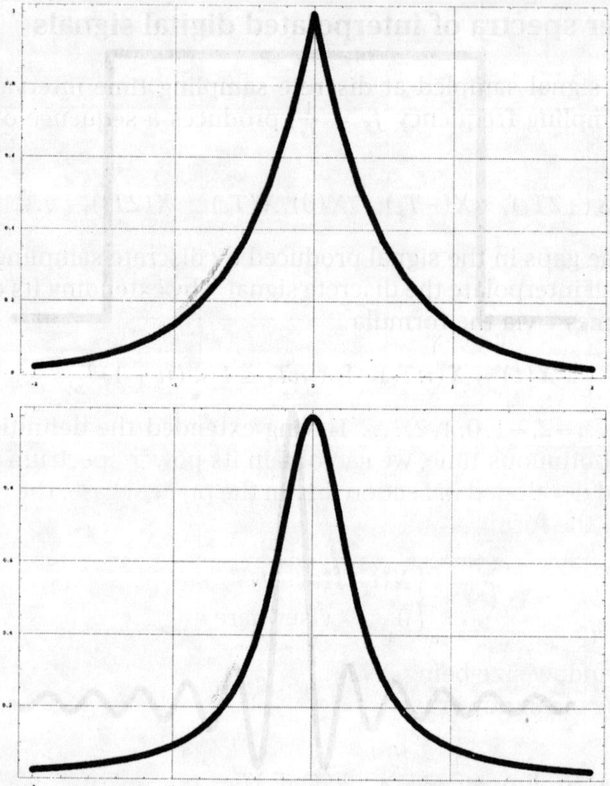

Fig. 5.2.3. *Top*: Autocorrelation function of the random switching signal from Example 5.2.5. *Bottom*: The corresponding power spectrum.

$$\gamma_X(\tau) = e^{-2|\tau|}.$$

Thus its power spectral density can be directly calculated by taking the Fourier transform of the autocorrelation function:

$$\begin{aligned} S_X(f) &= \int_{-\infty}^{\infty} e^{-2|t|} e^{-j2\pi f t}\,dt = \int_0^{\infty} e^{-(2+j2\pi f)t}\,dt + \int_{-\infty}^{0} e^{-(-2+j2\pi f)t}\,dt \\ &= \frac{1}{2}\frac{1}{1+j\pi f} + \frac{1}{2}\frac{1}{1-j\pi f} = \frac{1}{1+(\pi f)^2}. \end{aligned}$$

Observe that the autocorrelation function decays here exponentially as the time lag increases while the power spectrum decays only like the inverse square of the frequency when the latter goes to infinity. The situation is pictured in Figure 5.2.3.

5.3 Power spectra of interpolated digital signals

A random signal sampled at discrete sampling time interval T_s, that is, with sampling frequency $f_s = \frac{1}{T_s}$, produces a sequence of random quantities

$$\ldots, \quad X(-2T_s), \quad X(-T_s), \quad X(0), X(T_s), \quad X(2T_s), \quad \ldots. \tag{5.3.1}$$

To fill in the gaps in the signal produced by discrete sampling at times nT_s we shall interpolate the discrete signal[30] by extending its definition to other times t via the formula

$$X(t) = X(nT_s) \quad \text{for } nT_s \le t < (n+1)T_s, \tag{5.3.2}$$

and $n = \ldots, -2, -1, 0, 1, 2, \ldots$. Having extended the definition of the signal to continuous time, we can obtain its power spectrum following the method developed in Section 5.1. In the present case, the windowed signal is of the form

$$X_N(t) = \begin{cases} X(t) & \text{for } -NT_s \le t < NT_s; \\ 0 & \text{elsewhere,} \end{cases}$$

with the window size being $2NT_S$.

Now the mean power is

$$\begin{aligned} \mathbf{E}(\mathrm{PW}_X) &= \mathbf{E} \lim_{N\to\infty} \frac{1}{2NT_s} \sum_{n=-N}^{N-1} X^2(nT_s)T_s \\ &= \mathbf{E} \lim_{N\to\infty} \frac{1}{2NT_s} \int_{-\infty}^{\infty} |X_N(f)|^2 df = \int_{-\infty}^{\infty} \lim_{N\to\infty} \frac{\mathbf{E}|X_N(f)|^2}{2NT_s} df \\ &= \int_{-\infty}^{\infty} S(f)df, \end{aligned} \tag{5.3.3}$$

with the power spectral density

$$S(f) = \lim_{N\to\infty} \frac{\mathbf{E}|X_N(f)|^2}{2NT_s}, \tag{5.3.4}$$

and the equality in (5.3.3) resulting from the Parseval formula.

In the next step, we evaluate the Fourier transform $X_N(f)$ of the windowed interpolated signal which is needed in formula (5.3.4):

$$X_N(f) = \int_{-\infty}^{\infty} X_N(t)e^{-j2\pi ft}dt = \sum_{n=-N}^{N-1} \int_{nT_s}^{(n+1)T_s} X(t)e^{-j2\pi ft}dt$$

---

30 The material of this section should be compared with the analysis of the discrete and the fast Fourier transforms carried out in Section 2.7 for non-random, deterministic signals.

$$= \frac{1}{-j2\pi f} \sum_{n=-N}^{N-1} X(nT_s)(e^{-j2\pi f(n+1)T_s} - e^{-j2\pi f nT_s})$$

$$= \frac{1 - e^{-j2\pi f T_s}}{j2\pi f} \sum_{n=-N}^{N-1} X(nT_s)e^{-j2\pi f nT_s}.$$

Substituting this result into (5.3.4), we get the following structure of the power spectrum of $X(t)$:

$$S(f) = \lim_{N\to\infty} \frac{|1 - e^{-j2\pi f T_s}|^2}{4\pi^2 f^2} \cdot \frac{\mathbf{E}|\sum_{n=-N}^{N-1} X(nT_s)e^{-2\pi j f nT_s}|^2}{2NT_s}$$

$$= \frac{1 - \cos 2\pi f T_s}{2\pi^2 f^2} \lim_{N\to\infty} \sum_{k=-N}^{N-1} \sum_{n=-N}^{N-1} \gamma_X((n-k)T_s)e^{-2\pi j(n-k)fT_s} \frac{1}{2NT_s}.$$

Changing the second summation variable by substituting $n = m + k$, we get

$$S(f) = \frac{1 - \cos 2\pi f T_s}{2\pi^2 f^2} \lim_{N\to\infty} \sum_{k=-N}^{N-1} \sum_{m=-N-k}^{N-1-k} \gamma_X(mT_s)e^{-j2\pi m f T_s} \frac{1}{2NT_s}$$

$$= \frac{1 - \cos 2\pi f T_s}{2\pi^2 f^2 T_s^2} \cdot \sum_{m=-\infty}^{\infty} \gamma_X(mT_s)e^{-j2\pi m f T_s} T_s.$$

Hence the power spectrum can be written as a product

$$S(f) = S_1(f)S_2(f), \tag{5.3.5}$$

where the factor

$$S_1(f) = \frac{1 - \cos 2\pi f T_s}{2\pi^2 f^2 T_s^2} \tag{5.3.6}$$

decays to 0 at infinite frequencies ($f \to \pm\infty$) and is indepedent of the statistical properties of the signal (that is, of the autocorrelation function $\gamma_X(nT_s)$). The second factor,

$$S_2(f) = \sum_{m=-\infty}^{\infty} \gamma_X(mT_s)e^{-j2\pi m f T_s} T_s, \tag{5.3.7}$$

is a periodic function with period $f_s = \frac{1}{T_s}$, represented by the Fourier series with coefficients given by the discrete-time autocorrelation function of the discretely sampled signal.

Therefore, if instead of the original power spectrum we consider the ratio $\frac{S(f)}{S_1(f)}$, then we obtain clean relationships paralleling the symmetry of formulas for continuous-time signals:

$$\frac{S(f)}{S_1(f)} = \sum_{m=-\infty}^{\infty} \gamma_X(mT_s)e^{-j2\pi m f T_s} T_s \tag{5.3.8}$$

and

$$\gamma_X(mT_s) = \int_{-f_s/2}^{f_s/2} \frac{S(f)}{S_1(f)} e^{j2\pi mfT_s}\, df. \tag{5.3.9}$$

Remark 5.3.1. It is clear that all the relevant information about the spectrum of the signal sampled with the sampling interval T_s is contained in the frequency interval $(-\frac{f_s}{2}, +\frac{f_s}{2})$. Power assigned to higher frequencies, appearing in the side "lobes" of the spectrum (see Figure 5.3.1) is simply an artifact of the interpolation. Should we select a different interpolation scheme, the factor $S_1(f)$ responsible for the decay of the "lobes" would look different (see Section 5.4).

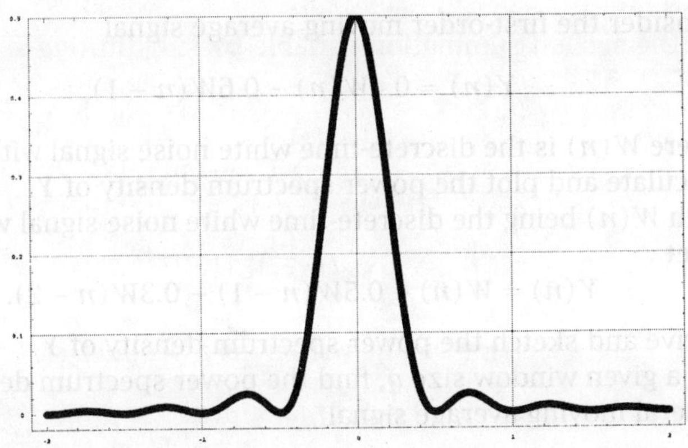

Fig. 5.3.1. Power spectrum of the interpolated moving average of the discrete-time white noise signal. The sampling rate is $f_s = \frac{1}{T_s} = 1$, and the relevant spectrum is concentrated in the interval $(-\frac{f_s}{2}, +\frac{f_s}{2})$. The side "lobes" are an artifact of the interpolation scheme.

Example 5.3.1 (interpolated moving average of the discrete-time white noise). Let the sampling interval $T_s = 1$, and let $W(n)$ be a discrete-time white noise signal ($EW(n) = 0$, $\gamma_W(n) = \frac{\delta(n)}{2}$). For the moving average signal

$$Y(n) = \frac{1}{2}W(n) + \frac{1}{2}W(n-1),$$

we have calculated in Chapter 4 that

$$\gamma_Y(0) = \frac{1}{4}, \qquad \gamma_Y(\pm 1) = \frac{1}{8}, \qquad \gamma_Y(k) = 0 \quad \text{for } |k| \geq 2.$$

Thus the periodic $S_2(f)$ factor of the power spectrum of the interpolated $Y(n)$ is of the form

$$S_2(f) = \frac{1}{8}e^{j2\pi f\cdot 1} + \frac{1}{4} + \frac{1}{8}e^{-j2\pi f\cdot 1} = \frac{1}{4}(1+\cos 2\pi f),$$

and the power spectral density itself of the interpolated Y is

$$S_Y(f) = S_1(f)S_2(f) = \frac{1-\cos 2\pi f}{2\pi^2 f^2}\cdot\frac{1}{4}(1+\cos 2\pi f)$$
$$= \frac{1}{2}\left[\frac{\sin(2\pi f)}{2\pi f}\right]^2.$$

5.4 Problems and exercises

In the first three problems, follow Example 5.3.1.

5.4.1. Consider the first-order moving average signal

$$Y(n) = 0.4W(n) - 0.6W(n-1),$$

where $W(n)$ is the discrete-time white noise signal with $\sigma_W^2 = 1$. Calculate and plot the power spectrum density of Y.

5.4.2. With $W(n)$ being the discrete-time white noise signal with $\sigma_W^2 = 5$, let

$$Y(n) = W(n) + 0.5W(n-1) - 0.3W(n-2).$$

Derive and sketch the power spectrum density of Y.

5.4.3. For a given window size q, find the power spectrum density of a general moving average signal

$$Y(n) = b_0W(n) + b_1W(n-1) + \cdots + b_qW(n-q),$$

where $W(n)$ is the discrete-time white noise with $\sigma_W^2 = 1$.

5.4.4. *Discrete sampling with linear interpolation.* Consider a signal X sampled at sampling interval T_s. Its interpolation to continuous-time signal is given by the following formula:

$$X(t) = \sum_{m=-\infty}^{\infty} X(mT_s)\Lambda(t-mT_s),$$

where the interpolating kernel is

$$\Lambda(t) = \begin{cases} 1-\frac{t}{T_s} & \text{for } 0 < t < T_s; \\ 1+\frac{t}{T_s} & \text{for } -T_s < t < 0; \\ 0 & \text{elsewhere.} \end{cases}$$

(a) Plot the kernel $\Lambda(t)$ and the interpolated $X(t)$ for an example of the sampled signal selected by you. Explain the interpolation effect.

(b) Demonstrate that the Fourier transform of the interpolated signal is of the form

$$X_N(f) = \sum_{m=-N}^{N} X(mT_s)e^{-2\pi jmT_s f}\Lambda(f),$$

where $\Lambda(f)$ is the Fourier transform of the kernel $\Lambda(t)$. Produce a plot of $\Lambda(f)$.

(c) Verify that the power spectrum density for the interpolated signal $X(t)$ is

$$S(f) = \lim_{N\to\infty} \frac{E|X_N(f)|^2}{(2N+1)T_s} = \Lambda^2(f)\frac{1}{T_s}\sum_{m=-\infty}^{\infty} \gamma_X(mT_s)e^{-2\pi jmfT_s}.$$

5.4.5. A stationary signal $X(t)$ has the autocorrelation function

$$\gamma_X(\tau) = 16e^{-5|\tau|}\cos 20\pi\tau + 8\cos 10\pi\tau.$$

(a) Find the variance of this signal.
(b) Find the power spectrum density of this signal.
(c) Find the value of the spectral density at zero frequency.

5.4.6. A stationary signal $X(t)$ has the spectral density of the form

$$S_X(f) = \begin{cases} 5 & \text{for } \frac{10}{2\pi} \le |f| \le \frac{20}{2\pi}; \\ 0 & \text{elsewhere.} \end{cases}$$

(a) Find the mean power of X.
(b) Find the autocorrelation function of X.
(c) Find the value of the autocorrelation at $\tau = 0$.

5.4.7. A stationary signal $X(t)$ has the spectral density of the form

$$S_X(f) = \frac{9}{(2\pi f)^2 + 64}.$$

At what frequency does the spectral density fall to one-half of its maximal value? (This value is called the *half-power bandwidth.*)

(a) Write an expression for the spectral density of a band-limited white noise Y that has the same value at zero frequency and the same mean power as X. What is its bandwidth? It is called the *equivalent-noise bandwidth* of X. Compare it with the half-power bandwidth.
(b) Find the autocorrelation function of signal X.
(c) Find the autocorrelation function of signal Y.
(d) Compare the values of these two autocorrelation functions at $\tau = 0$.

5.4.8. Use formula (5.2.2) for the periodogram to estimate spectra of signals simulated in Figures 4.1.3, 4.1.5, and 4.1.6 and Problem 4.3.6. Then check your results by superimposing the periodograms on the corresponding theoretical spectral densities. You may want to eliminate some noise from the periodograms by using the moving average technique applied in Chapter 4 to random signals themselves.

6

Transmission of Stationary Signals through Linear Systems

Signals produced in nature are almost never experienced in their original form. Usually, we have access to them after they pass through various sensing and/or transmission devices such as a voltmeter, for electric signals, the ear, for acoustic signals, the eye, for visual signals, a fiber optic cable, for wide-band Internet signals, etc. All of them impose restrictions on the signal being transmitted by attenuating different frequency components of the signal to a different degree. This process is generally called *filtering* and the devices that change the signal's spectrum are traditionally called *filters*.

A typical example here is the so-called *band-pass filter*, which permits transmission of the components of the signal only in a certain frequency band, attenuating the frequencies in that band in a uniform fashion, but totally "killing" the frequencies outside this band. Figure 6.0.1 shows results of filtering a portion of the EEG signal from Figure 4.1.1 through four band-pass filters with frequency bands (top to bottom) 0.5–3.5 Hz, 4–7.5 Hz, 8–12.5 Hz, and 13–17 Hz. In neurological literature the contents of the EEG signal within these frequency bands are traditionally called delta, theta, alpha, and beta waves, respectively.

In this chapter we study how statistical characteristics of random stationary signals are affected by transmission through linear filters. The linearity assumption means that we suppose that there is a linear relationship between the signals on the input and on the output of the filter. In real life this is not always the case, but the study of nonlinear filters is much more difficult than the linear theory presented below, and beyond the scope of this book.

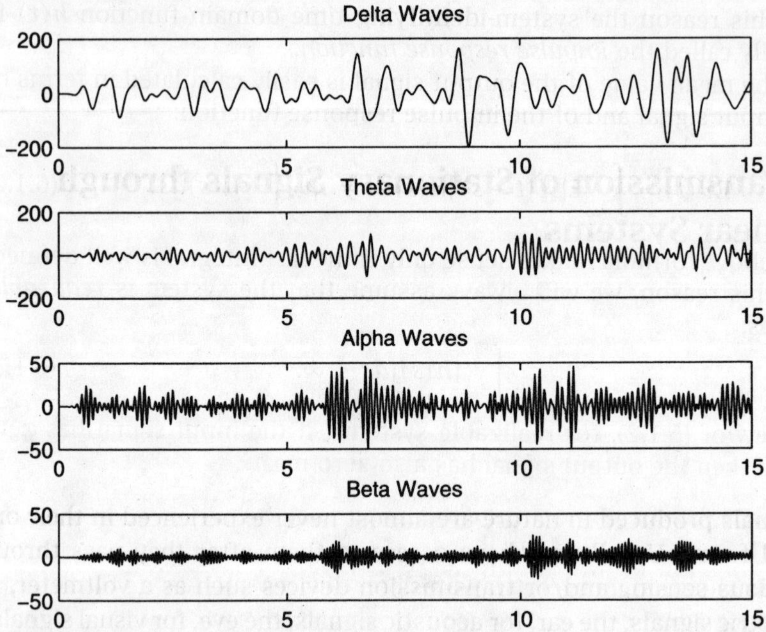

Fig. 6.0.1. A portion of the EEG signal from Figure 4.1.1 filtered through four band-pass filters with frequency bands (top to bottom) 0.5–3.5 Hz, 4–7.5 Hz, 8–12.5 Hz, and 13 – –17 Hz, respectively.

6.1 The time domain analysis

In this section we conduct the time domain analysis of transmission of random signals through a linear system shown schematically below:

$$X(t) \longrightarrow \boxed{h(t)} \longrightarrow Y(t).$$

The input signal $X(t)$ is assumed to be random and stationary with mean $m_X = \mathbf{E}X(t)$ and autocorrelation function $\gamma_X(\tau) = \mathbf{E}X(t)X(t+\tau)$. The system is identified by a function $h(t)$, and the output signal $Y(t)$ is defined as the continuous-time moving average (convolution):

$$Y(t) = \int_{-\infty}^{\infty} X(s)h(t-s)\,ds = \int_{-\infty}^{\infty} X(t-s)h(s)\,ds. \tag{6.1.1}$$

Note that in the case of a nonrandom Dirac delta impulse input $\delta(t)$ the nonrandom output signal is

$$y(t) = \int_{-\infty}^{\infty} \delta(s)h(t-s)\,ds = h(t-0) = h(t).$$

For this reason the system-identifying time domain function $h(t)$ is usually called the *impulse response function.*

The mean value of the output signal is easily calculated in terms of the input signal and of the impulse response function:

$$\mathbf{E}Y(t) = \int_{-\infty}^{\infty} \mathbf{E}[X(t-s)]h(s)ds = m_X \int_{-\infty}^{\infty} h(s)ds. \tag{6.1.2}$$

The above formula makes sense only if the last integral is well defined. For this reason, we will always assume that the system is *realizable,* that is,

$$\int_{-\infty}^{\infty} |h(s)|ds < \infty. \tag{6.1.3}$$

In view of (6.1.2), for realizable systems, if the input signal has zero mean then the output signal has also zero mean:

$$m_X = 0 \Longrightarrow m_Y = 0.$$

In this situation, henceforth we will restrict our attention only to zero-mean signals.

The calculation of the autocorrelation function of the output signal $Y(t)$ is a little bit more involved. Replacing the product of the integrals by the double integral, we obtain that

$$\begin{aligned}\gamma_Y(\tau) &= \mathbf{E}(Y(t)Y(t+\tau)) \\ &= \mathbf{E}\left[\int_{-\infty}^{\infty} X(t-s)h(s)ds \int_{-\infty}^{\infty} X(t+\tau-u)h(u)du\right] \\ &= \int_{-\infty}^{\infty}\int_{-\infty}^{\infty} \mathbf{E}[X(t-s)X(t+\tau-u)]h(s)h(u)dsdu.\end{aligned}$$

Then in view of the stationarity assumption,

$$\mathbf{E}[X(t-s)X(t+\tau-u)] = \mathbf{E}[X(-s)X(\tau-u)] = \gamma_X(\tau-u+s),$$

so that, finally,

$$\gamma_Y(\tau) = \int_{-\infty}^{\infty}\int_{-\infty}^{\infty} \gamma_X(\tau-u+s)h(s)h(u)dsdu. \tag{6.1.4}$$

A system is said to be *causal* if the current values of the output depend only on the past and present values of the input. This property can be equivalently stated as the requirement that the impulse response function satisfy

$$h(t) = 0 \quad \text{for } t \le 0. \tag{6.1.5}$$

In other words, the moving average is performed only over the past. This condition, in particular, implies that the second output integral in (6.1.1) is restricted to the positive half-line

$$Y(t) = \int_0^\infty X(t-s)h(s)ds, \tag{6.1.6}$$

and the autocorrelation function formula (6.1.4) becomes

$$\gamma_Y(\tau) = \int_0^\infty \int_0^\infty \gamma_X(\tau - u + s)h(s)h(u)dsdu. \tag{6.1.7}$$

In what follows, we will just consider causal systems.

Example 6.1.1 (an integrating circuit). A standard integrating circuit with a single capacitor is shown in Figure 6.1.1.

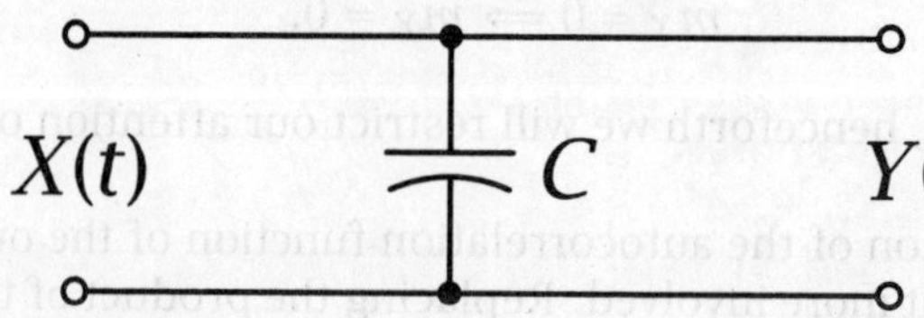

Fig. 6.1.1. A standard integrating circuit. The voltage $Y(t)$ on the output is the integral of the current $X(t)$ on the input.

The impulse response function for this system is the unit step function $u(t)$ multiplied by $\frac{1}{C}$, where the constant C represents the capacitance of the capacitor:

$$h(s) = \frac{1}{C}u(s) = \begin{cases} 0 & \text{for } s < 0; \\ \frac{1}{C} & \text{for } s \geq 0. \end{cases}$$

The output is

$$Y(t) = \frac{1}{C}\int_{-\infty}^{\infty} X(s)U(t-s)ds = \frac{1}{C}\int_{-\infty}^{t} X(s)ds.$$

Obviously, this system, although causal, is not realizable since

$$\int_{-\infty}^{\infty} |h(t)|dt = \int_0^\infty \frac{1}{C}dt = \infty.$$

To avoid this difficulty, we need to restrict the integrating circuit to a finite time interval and assume that the adjusted impulse response function is of the form

$$h(s) = \begin{cases} 0 & \text{for } s < 0; \\ \frac{1}{C} & \text{for } 0 \le s \le T; \\ 0 & \text{for } s > T. \end{cases} \tag{6.1.8}$$

In this situation, the system is realizable and the output is

$$Y(t) = \int_{-\infty}^{\infty} X(s)h(t-s)ds = \frac{1}{C}\int_{t-T}^{t} X(s)ds.$$

The autocorrelation is

$$\begin{aligned} \gamma_Y(\tau) &= \int_0^T \int_0^T \gamma_X(\tau - u + s)h(s)h(u)dsdu \\ &= \frac{1}{C^2}\int_0^T \int_0^T \gamma_X(u - (\tau + s))dsdu \end{aligned} \tag{6.1.9}$$

because of the symmetry of the autocorrelation function.

Therefore, if the input signal is the standard white noise $X(t) = W(t)$ with the autocorrelation $\gamma_W(t) = \delta(t)$, then for $\tau \ge 0$, the output autocorrelation function is

$$\gamma_Y(\tau) = \int_0^T \int_0^T \delta(u - (\tau + s))duds = \int_0^T \zeta(s)ds,$$

where

$$\zeta(s) = \begin{cases} 0 & \text{for } \tau + s < 0; \\ \frac{1}{2} & \text{for } \tau + s = 0; \\ 1 & \text{for } 0 < \tau + s < T; \\ \frac{1}{2} & \text{for } \tau + s = T; \\ 0 & \text{for } \tau + s > T. \end{cases}$$

Hence

$$\gamma_Y(s) = \begin{cases} 0 & \text{for } \tau < -T; \\ T - |\tau| & \text{for } -T \le \tau \le T; \\ 0 & \text{for } \tau > T. \end{cases} \tag{6.1.10}$$

If the input signal $X(t)$ is a simple random harmonic oscillation with the autocorrelation function $\gamma_X(\tau) = \cos\tau$, then the output autocorrelation is

$$\gamma_Y(\tau) = \int_0^{\infty}\int_0^{\infty} \cos(\tau - u + s)dsdu = -\cos(\tau + T) + 2\cos\tau - \cos(\tau - T). \tag{6.1.11}$$

As simple as the formula (6.1.9) for the output autocorrelation function seems to be, the analytic evaluation of the double convolution may get tedious very quickly. Consider, for example, an input signal $X(t)$ with the autocorrelation function

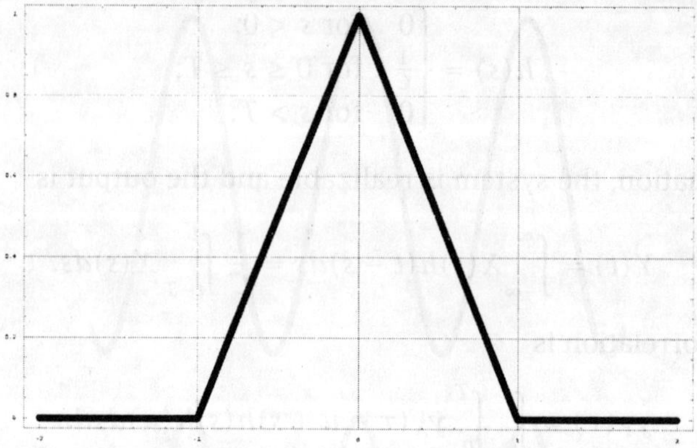

Fig. 6.1.2. The output autocorrelation function $\gamma_Y(\tau)$ (6.1.10) of the integrating system (6.1.8) with $T = 1$, in the case of the standard white noise input $X(t) = W(t)$.

$$\gamma_X(\tau) = \frac{1}{1+\tau^2}, \tag{6.1.12}$$

which corresponds to the exponentially decaying power spectrum (see Section 6.4).

In this case,

$$\begin{aligned}\gamma_Y(\tau) &= \int_0^T \int_0^T \frac{1}{1+(\tau-u+s)^2}\,ds\,du \qquad (6.1.13)\\ &= \frac{1}{2}(2(T-\tau)\arctan(T-\tau) - 2\tau\arctan\tau - \log(1+(T-\tau)^2)\\ &\qquad + \log(1+\tau^2))\\ &\quad + \frac{1}{2}(-2\tau\arctan(\tau) + 2(\tau+T)\arctan(\tau+T) + \log(1+\tau^2)\\ &\qquad + \log(1+T^2+2T\tau+\tau^2)).\end{aligned}$$

So even for a relatively simple input autocorrelation function the output autocorrelation is quite complex and unreadable. Yes, you guessed right—we have obtained this formula using *Mathematica.* Figure 6.1.4 traces graphically the dependence of $\gamma_Y(\tau)$ on T.

Example 6.1.2 (an RC filter). A standard RC filter is shown in Figure 6.1.5.

The impulse response function of this circuit is of the form

$$h(t) = \frac{1}{RC}\exp\left(-\frac{t}{RC}\right)\cdot u(t), \tag{6.1.14}$$

where $u(t)$ is the unit step function, R is the resistivity, and C is the capacitance. The product RC represent the so-called time constant of the circuit.

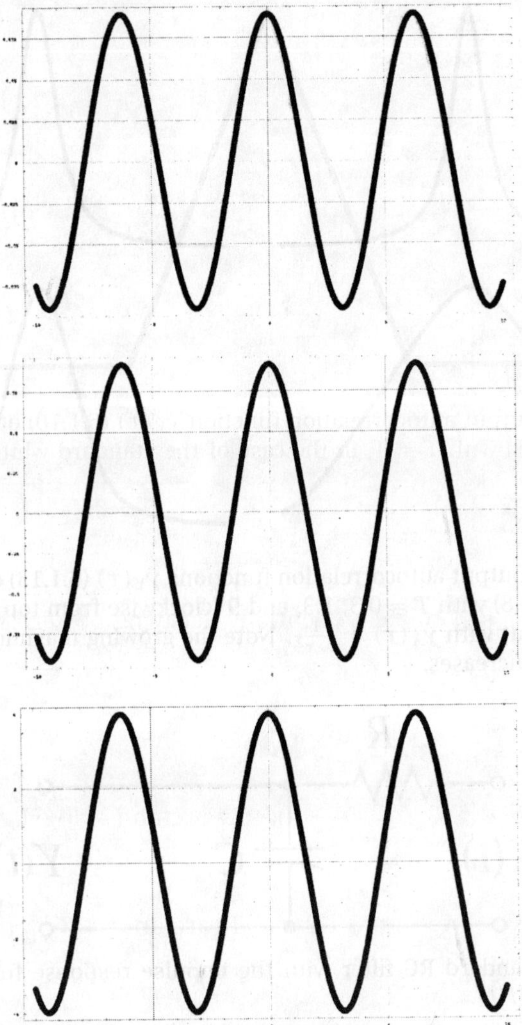

Fig. 6.1.3. The output autocorrelation functions $\gamma_Y(\tau)$ (6.1.11) of the integrating system (6.1.8) with $T = 0.3$, 1, and 3 (top to bottom) in the case of a simple random harmonic oscillation input with $\gamma_X(\tau) = \cos\tau$. Note the increasing amplitude of $\gamma_Y(\tau)$ as T increases.

In the case of the white noise input signal with $\gamma_X(\tau) = \delta(\tau)$, the output autocorrelation function, for $\tau > 0$, is

$$\gamma_Y(\tau) = \int_0^\infty \int_0^\infty \delta(u - (s+\tau))h(u)h(s)duds = \int_0^\infty h(s+\tau)h(s)ds$$
$$= \int_0^\infty \frac{1}{RC}e^{s+\tau/RC} \cdot \frac{1}{RC}e^{s/RC}ds = \frac{1}{2RC}e^{-\tau/RC}.$$

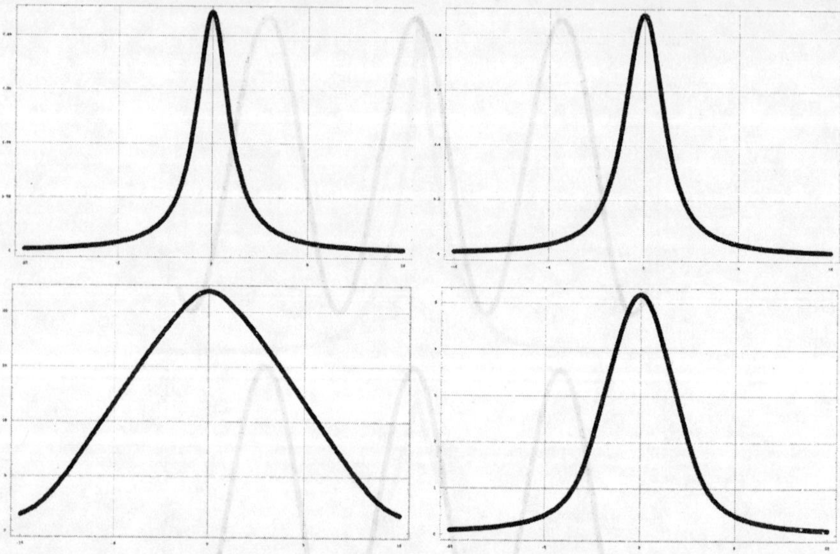

Fig. 6.1.4. The output autocorrelation functions $\gamma_Y(\tau)$ (6.1.13) of the integrating system (6.1.8) with $T = 0.3$, 1.3, and 9 (clockwise from top left corner), in the case of input with $\gamma_X(\tau) = \frac{1}{1+\tau^2}$. Note the growing maximum and spread of $\gamma_Y(\tau)$ as T increases.

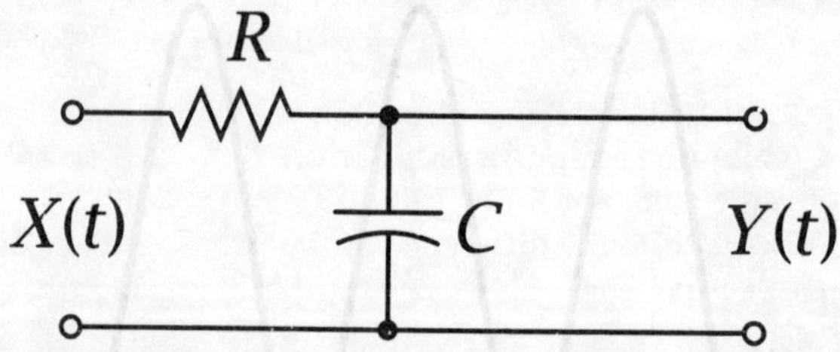

Fig. 6.1.5. A standard RC filter with the impulse response function $h(t) = \frac{1}{RC}\exp(-\frac{t}{RC}) \cdot U(t)$.

So

$$\gamma_Y(\tau) = \frac{1}{2RC}\exp\left(-\frac{|\tau|}{RC}\right). \tag{6.1.15}$$

The shape of the output autocorrelation function for small and large values of the RC constant is shown in Figure 6.1.6.

For the simple random harmonic oscillation with autocorrelation $\gamma_X(\tau) = \cos\tau$ as the input, the output autocorrelation is

$$\begin{aligned}\gamma_Y(\tau) &= \int_0^\infty \int_0^\infty \cos(\tau - u + s)\frac{1}{RC}\exp\left(\frac{-s}{RC}\right)\frac{1}{RC}\exp\left(-\frac{u}{RC}\right) ds\,du \\ &= \frac{\cos\tau}{1+(RC)^2}.\end{aligned}$$

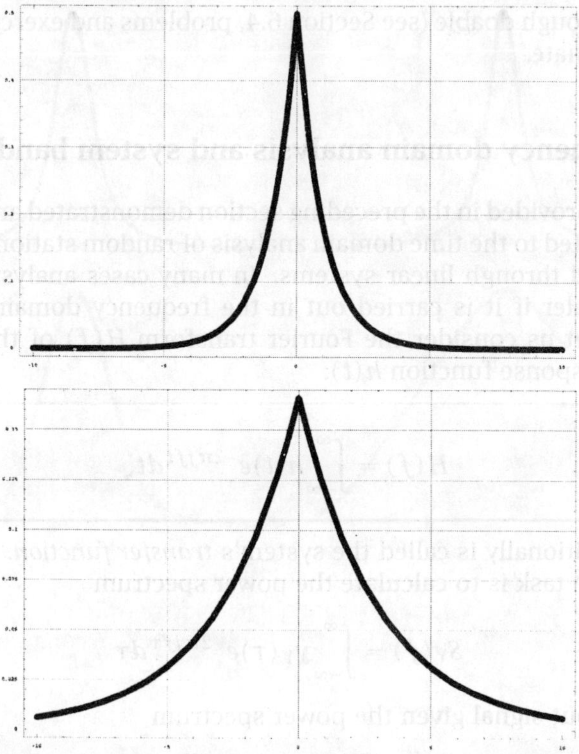

Fig. 6.1.6. The output autocorrelation function $\gamma_Y(\tau)$ for the RC filter (6.1.14) with a standard white noise input with $\gamma_X(\tau) = \delta(\tau)$. The top figure shows the case of small time constant $RC = 1$ and the bottom the case of the larger time constant $RC = 3$. Note the difference in the maximum and the spread of $\gamma_Y(\tau)$ in these two cases.

But a slightly more complex input autocorrelation function

$$\gamma_X(\tau) = e^{-2|\tau|},$$

corresponding to the switching input signal, produces the output autocorrelation function of the form

$$\begin{aligned}\gamma_Y(\tau) &= \frac{1}{(RC)^2}\int_0^\infty\int_0^\infty e^{-|\tau-u+s|}e^{-(s+u)/(RC)}\,ds\,du \qquad (6.1.16)\\ &= \frac{1}{(RC)^2}\left[\int_0^\tau\int_0^\infty e^{-(\tau-u+s)}e^{-(s+u)/(RC)}\,ds\,du\right.\\ &\quad+\int_\tau^\infty\left(\int_0^{u-\tau} e^{\tau-u+s}e^{-(s+u)/(RC)}\,ds\right.\\ &\quad\left.\left.+\int_{u-\tau}^\infty e^{-(\tau-u+s)}e^{-(s+u)/(RC)}\,ds\right)du\right]\end{aligned}$$

which, although doable (see Section 6.4, problems and exercises), is not fun to evaluate.

6.2 Frequency domain analysis and system bandwidth

Examples provided in the preceding section demonstrated analytic difficulties related to the time domain analysis of random stationary signals transmitted through linear systems. In many cases analysis becomes much simpler if it is carried out in the frequency domain. For this purpose, let us consider the Fourier transform $H(f)$ of the system's impulse response function $h(t)$:

$$H(f) = \int_{-\infty}^{\infty} h(t)e^{-2\pi jft}dt, \tag{6.2.1}$$

which traditionally is called the system's *transfer function*.

Now the task is to calculate the power spectrum

$$S_Y(f) = \int_{-\infty}^{\infty} \gamma_Y(\tau)e^{-2\pi jf\tau}d\tau \tag{6.2.2}$$

of the output signal given the power spectrum

$$S_X(f) = \int_{-\infty}^{\infty} \gamma_X(\tau)e^{-2\pi jf\tau}d\tau$$

of the input signal. Since the output autocorrelation function $\gamma_Y(t)$ has been calculated in Section 6.1, substituting the expression obtained in (6.1.4) into (6.2.1), we get

$$\begin{aligned} S_Y(f) &= \int_{-\infty}^{\infty}\left(\int_{-\infty}^{\infty}\int_{-\infty}^{\infty} \gamma_X(\tau - s + u)h(s)h(u)dsdu\right)e^{-2\pi jf\tau}d\tau \\ &= \int_{-\infty}^{\infty}\int_{-\infty}^{\infty}\left(\int_{-\infty}^{\infty} \gamma_X(\tau - s + u)e^{-2\pi jf(\tau-s+u)}d\tau\right)h(s)e^{-2\pi jfs}ds \\ &\cdot h(u)e^{2\pi jfu}du. \end{aligned}$$

Making the substitution $\tau - s + u = w$ in the inner integral, we arrive at the final formula

$$S_Y(f) = S_X(f) \cdot H(f) \cdot H^*(f) = S_X(f) \cdot |H(f)|^2. \tag{6.2.3}$$

So the output power spectrum is obtained simply by multiplying the input power spectrum by a fixed factor $|H(f)|^2$, which is called the system's *power transfer function*.

The appearance of power transfer function $|H(f)|^2$ in formula (6.2.3) suggests introduction of the concept of the system's bandwidth. As in the case of signals (see Section 5.4) several choices are possible.

The *equivalent-noise bandwidth* BW_n is defined as the cutoff frequency $f_{\max}$ of the limited-band white noise with the amplitude equal to the value of the system's power transfer function at 0 and the mean power equal to the integral of the system's power transfer function, that is,

$$2\,\mathrm{BW}_n\,|H(0)|^2 = \int_{-\infty}^{\infty} |H(f)|^2 df,$$

which gives

$$\mathrm{BW}_n = \frac{1}{2|H(0)|^2}\int_{-\infty}^{\infty} |H(f)|^2 df. \tag{6.2.4}$$

The *half-power bandwidth* $\mathrm{BW}_{1/2}$ is defined as the frequency where the system's power transfer function declines to one-half of its maximum value which is always equal to $|H(0)|^2$. Thus it is obtained by solving, for $\mathrm{BW}_{1/2}$, the equation

$$|H(\mathrm{BW}_{1/2})|^2 = \frac{1}{2}|H(0)|^2. \tag{6.2.5}$$

Obviously, the above bandwidth concepts make the most sense for lowpass filters, that is, in the case when the system's power transfer function has a distinctive maximum at 0, dominating its values elsewhere. But for other systems such as band-pass filters, similar bandwidth definitions can be easily devised.

Example 6.2.1 (an RC filter). Recall that in this case the impulse response function is given by

$$h(t) = \frac{1}{RC}e^{-\frac{t}{RC}} \cdot u(t).$$

So the transfer function is

$$H(f) = \int_{-\infty}^{\infty} h(t)e^{-2\pi jft}dt = \int_0^{\infty} \frac{1}{RC}e^{-\frac{t}{RC}}e^{-2\pi jft}dt = \frac{1}{1+2\pi jRCf},$$

and, consequently, the power transfer function is

$$|H(f)|^2 = \frac{1}{1+2\pi jRCf}\cdot\frac{1}{1-2\pi jRCf} = \frac{1}{1+(2\pi RCf)^2}. \tag{6.2.6}$$

The half-power bandwidth of the RC filter is easily computable from the equation

$$\frac{1}{1+(2\pi RC(\mathrm{BW}_{1/2}))^2} = \frac{1}{2},$$

which gives

$$BW_{1/2} = \frac{1}{2\pi RC}.$$

The bandwidth decreases hyperbolically with the increase of the RC constant.

The output power spectra for an RC filter are thus easily evaluated. In the case of the standard white noise input with $S_X(f) \equiv 1$, the output power spectrum is

$$S_Y(f) = \frac{1}{1 + (2\pi RCf)^2}.$$

If the input signal is a random oscillation with the power spectrum

$$S_X(f) = \frac{A_0^2}{2}(\delta(f - f_0) + \delta(f + f_0)),$$

then the output power spectrum is

$$S_Y(f) = \frac{A_0^2}{2}(\delta(f - f_0) + \delta(f + f_0)) \cdot \frac{1}{1 + (2\pi RCf)^2}.$$

If the input is a switching signal with the power spectrum

$$S_X(f) = \frac{1}{1 + (af)^2},$$

then the output power spectrum is

$$S_Y(f) = \frac{1}{1 + (af)^2} \cdot \frac{1}{1 + (2\pi RCf)^2}.$$

Example 6.2.2 (bandwidth of the finite-time integrating circuit). Let us calculate the bandwidths BW_n and $BW_{1/2}$ for the finite-time integrator with the impulse response function

$$h(t) = \begin{cases} 1 & \text{for } 0 \le t \le T; \\ 0 & \text{elsewhere.} \end{cases}$$

In this case, the transfer function is

$$H(f) = \int_0^T e^{-2\pi jft} dt = \frac{1}{2\pi jf}(1 - e^{-2\pi jfT}),$$

so that the power transfer function is

$$|H(f)|^2 = \frac{(1 - e^{-2\pi jfT})(1 - e^{2\pi jfT})}{(2\pi f)^2} = \frac{2(1 - \cos 2\pi fT)}{(2\pi f)^2}. \quad (6.2.7)$$

Finding directly the integral of the power transfer function is a little tedious, but fortunately, by Parseval's formula,

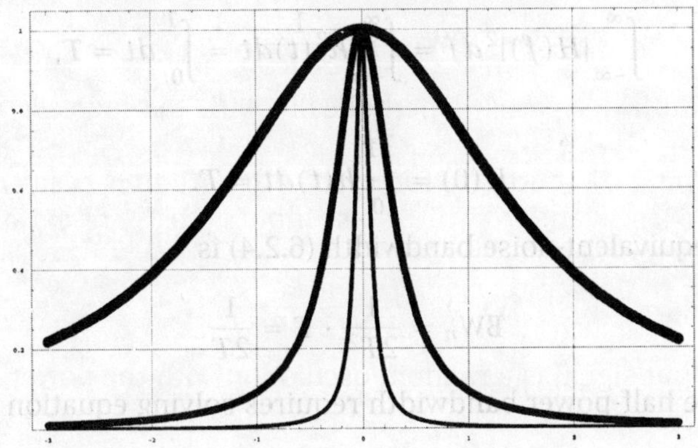

Fig. 6.2.1. Power transfer functions $|H(f)|^2 = \frac{1}{1+(2\pi RCf)^2}$ for the RC filter with the RC constants 0.1 (thick line), 0.5 (medium line), and 2.0 (thin line). The half-power bandwidths $\mathrm{BW}_{1/2}$ are, respectively, 1.6, 0.32, and 0.08.

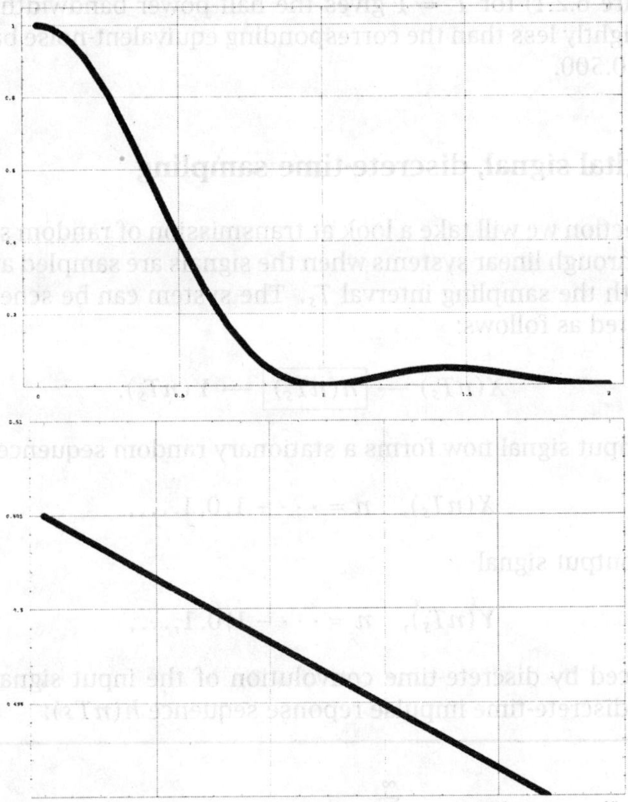

Fig. 6.2.2. *Top*: Power transfer function (6.2.7) of the finite-time integrating circuit with $T = 1$. *Bottom*: Magnified portion of the power transfer function for f between 0.44 and 0.45. This graphical analysis gives the half-power bandwidth $\mathrm{BW}_{1/2} = 0.443$.

$$\int_{-\infty}^{\infty} |H(f)|^2 df = \int_{-\infty}^{\infty} h^2(t)dt = \int_0^T dt = T,$$

and

$$H(0) = \int_0^T h(t)dt = T.$$

Thus the equivalent-noise bandwidth (6.2.4) is

$$\mathrm{BW}_n = \frac{1}{2T^2} \cdot T = \frac{1}{2T}.$$

Finding the half-power bandwidth requires solving equation (6.2.5):

$$\frac{2(1 - \cos 2\pi(\mathrm{BW}_{1/2})T)}{(2\pi(\mathrm{BW}_{1/2}))^2} = \frac{T^2}{2},$$

which can be done only numerically. Indeed, a quick graphical analysis (see Figure 6.2.1) for $T = 1$ gives the half-power bandwidth $\mathrm{BW}_{1/2} = 0.443$, slightly less than the corresponding equivalent-noise bandwidth $\mathrm{BW}_{\mathrm{eqn}} = 0.500$.

6.3 Digital signal, discrete-time sampling

In this section we will take a look at transmission of random stationary signals through linear systems when the signals are sampled at discrete times with the sampling interval T_s. The system can be schematically represented as follows:

$$X(nT_s) \longrightarrow \boxed{h(nT_s)} \longrightarrow Y(nT_s).$$

The input signal now forms a stationary random sequence

$$X(nT_s), \quad n = \cdots -1, 0, 1, \ldots, \tag{6.3.1}$$

and the output signal

$$Y(nT_s), \quad n = \cdots -1, 0, 1, \ldots, \tag{6.3.2}$$

is produced by discrete-time convolution of the input signal $X(nT_s)$ with the discrete-time impulse reponse sequence $h(nTs)$:

$$Y(nT_s) = \sum_{i=-\infty}^{\infty} X(iT_s)h(nT_s - iT_s)T_s. \tag{6.3.3}$$

In the discrete-time case, the realizability condition is

$$\sum_{n=-\infty}^{\infty} |h(nT_s)| < \infty,$$

and the causality condition means that

$$h(nT_s) = 0 \quad \text{for } n < 0.$$

With discrete-time inputs and outputs the autocorrelation functions are just discrete sequences and are defined by the formulas

$$\gamma_X(kT_s) = \mathrm{E}(X(nT_s)X(nT_s + kT_s)),$$
$$\gamma_Y(kT_s) = E(Y(nT_s)Y(nT_s + kT_s)).$$

Then after a direct application of (6.3.3), one obtains the following formula for the output autocorrelation sequence as a function of the input autocorrelation sequence and the impulse response sequence:

$$\gamma_Y(kT_s) = \sum_{l=-\infty}^{\infty} \sum_{i=-\infty}^{\infty} \gamma_X(kT_s - lT_s + iT_s)h(lT_s)h(iT_s)T_s^2. \tag{6.3.4}$$

To move into the frequency domain one can either directly apply the discrete or fast Fourier transforms or, as in Section 6.3, use the straight continuous-time Fourier transform technique assuming that both the signal and the impulse response function have been interpolated by constants between sampling points. We will follow the latter approach. Therefore, using formula (5.3.5), we get

$$S_X(f) = S_1(f) \cdot S_{2,X}(f), \tag{6.3.5}$$

with

$$S_{2,X}(f) = \sum_{m=-\infty}^{\infty} \gamma_X(mT_s)e^{-j2\pi mfT_s}T_s,$$

and

$$S_Y(f) = S_1(f) \cdot S_{2,Y}(f), \tag{6.3.6}$$

with

$$S_{2,Y}(f) = \sum_{m=-\infty}^{\infty} \gamma_Y(mT_s)e^{-j2\pi mfT_s}T_s,$$

and

$$S_1(f) = \frac{1 - \cos 2\pi f T_s}{2\pi^2 f^2 T_s^2}.$$

Remember that all the relevant information about the discrete sampled signal is contained in the frequency interval $(-\frac{f_s}{2}, \frac{f_s}{2})$ (see Remark 5.3.1). The transfer function of this system is

$$H(f) = \int_{-\infty}^{\infty} h(t)e^{-j2\pi ft}dt = \sum_{k=-\infty}^{\infty} h(kT_s) \int_{kT_s}^{(k+1)T_s} e^{-j2\pi ft}dt$$

$$= \frac{1-e^{j2\pi fT_s}}{-j2\pi fT_s} \sum_{k=-\infty}^{\infty} h(kT_s)e^{-j2\pi fkT_s}T_s, \tag{6.3.7}$$

so that the power transfer function

$$|H(f)|^2 = \frac{1-\cos 2\pi fT_s}{2\pi^2 f^2 T_s^2} \sum_{k=-\infty}^{\infty} \sum_{n=-\infty}^{\infty} h(kT_s)h(nT_s)e^{-j2\pi f(k-n)T_s}T_s^2. \tag{6.3.8}$$

Again, all the relevant information about the discrete power transfer function contained in the frequency interval $(-\frac{f_s}{2}, \frac{f_s}{2})$ (see Remark 5.3.1).

Finally, since we already know from Section 6.2 that

$$S_Y(f) = |H(f)|^2 S_X(f),$$

we also get from (6.3.5)–(6.3.6) that

$$S_{2,Y}(f) = |H(f)|^2 S_{2,X}(f). \tag{6.3.9}$$

or, equivalently,

$$\sum_{m=-\infty}^{\infty} \gamma_Y(mT_s)e^{-j2\pi mfT_s}T_s = |H(f)|^2 \cdot \sum_{m=-\infty}^{\infty} \gamma_X(mT_s)e^{-j2\pi mfT_s}T_s. \tag{6.3.10}$$

Example 6.3.1 (autoregressive moving average system (ARMA)). We now take the sampling period $T_s = 1$ and the output $Y(n)$ determined from the input $X(n)$ via the autoregressive moving average scheme with parameters p and q (in brief, ARMA(p,q)):

$$Y(n) = \sum_{l=0}^{q} b(l)X(n-l) - \sum_{l=1}^{p} a(l)Y(n-l). \tag{6.3.11}$$

Defining $a(0) = 1$, we can then write

$$\sum_{l=0}^{p} a(l)Y(n-l) = \sum_{l=0}^{q} b(l)X(n-l).$$

Since the Fourier transform of the convolution is a product of Fourier transforms, we have

$$X(f)\sum_{l=0}^{q} b(l)e^{-2\pi jflT} = Y(f)\sum_{l=0}^{p} a(l)e^{-2\pi jflT},$$

so the transfer function

$$H(f) = \frac{Y(f)}{X(f)} = \frac{\sum_{l=0}^{q} b(l)e^{-2\pi jflT}}{\sum_{l=0}^{p} a(l)e^{-2\pi jflT}}. \tag{6.3.12}$$

Example 6.3.2 (a solution of the stochastic difference equation). This example was considered in Chapter 4, but let us observe that it is a special case of Example 6.3.1, with parameters $p = 1$, $q = 0$, and the input signal being the standard discrete white noise $W(n)$ with $\sigma_W^2 = 1$. In other words,

$$Y(n) = -a_1 Y(n-1) + b_0 W(n).$$

In view of (6.3.12), the power transfer function is

$$|H(f)|^2 = \frac{b_0}{1 + a_1 e^{-2\pi j f}} \cdot \frac{b_0}{1 + a_1 s e^{2\pi j f}} = \frac{b_0^2}{1 + a_1^2 + 2a_1 \cos 2\pi f},$$

with, again, all the relevant information contained in the frequency interval $-\frac{1}{2} < f < \frac{1}{2}$.

Given that the input is the standard white noise, we have that

$$S_Y(f) = |H(f)|^2 \cdot 1 = \frac{b_0^2}{1 + a_1^2 + 2a_1 \cos 2\pi f}. \tag{6.3.13}$$

One way to find the output autocorrelation sequence $\gamma_Y(n)$ would be to take into account the relationship (6.3.10) and expand (6.3.13) into the Fourier series; its coefficients will form the desired autocorrelation sequence. This procedure is streightforward and requires only an application of the formula for the sum of a geometric series (see Section 6.4).

However, we would like to explore here a different route and employ a recursive procedure to find the output autocorrelation sequence. First, observe that

$$\begin{aligned}
\gamma_Y(k) &= \mathbf{E}(Y(n)Y(n+k)) \\
&= \mathbf{E}(-a_1 Y(n-1) + b_0 X(n)) \cdot (-a_1 Y(n+k-1) + b_0 X(n+k)) \\
&= a_1^2 \mathbf{E}(Y(n-1)Y(n+k-1)) - a_1 b_0 \mathbf{E}(Y(n-1)X(n+k))) \\
&\quad - a_1 b_0 \mathbf{E}(X(n)Y(n+k-1)) + b_0^2 \mathbf{E}(X(n)X(n+k)) \\
&= a_1^2 \gamma_Y(k) - a_1 b_0 \gamma_{XY}(k-1) + b_0)^2 \gamma_X(k),
\end{aligned}$$

where

$$\gamma_{XY}(k) = \mathbf{E}(X(n)Y(n+k)),$$

is the cross-correlation sequence of signals $X(n)$ and $Y(n)$. Thus

$$\gamma_Y(k) = \frac{b_0}{1 - a_1^2}(-a_1 \gamma_{XY}(k-1) + b_0 \gamma_X(k)).$$

For $k = 0$,

$$\gamma_Y(0) = \sigma_Y^2 = \frac{b_0}{1 - a_1^2}(-a_1 \mathbf{E}(X(n)Y(n-1)) + b_0 \gamma_X(0))$$

$$= \frac{b_0^2}{1-a_1^2}\gamma_X(0) = \frac{b_0^2}{1-a_1^2}.$$

For $k = 1$,

$$\gamma_Y(1) = \frac{b_0}{1-a_1^2}(-a_1\gamma_{XY}(0) + b_0\gamma_X(1)) = \frac{b_0(-a_1)}{1-a_1^2}\mathbf{E}(X(0)Y(0))$$

$$= \frac{b_0(-a_1)}{1-a_1^2}\mathbf{E}(X(0)(a_1Y(-1) + b_0X(0))) = \frac{b_0^2(-a_1)}{1-a_1^2}.$$

For a general $k > 1$,

$$\gamma_Y(k) = \frac{b_0}{1-a_1^2}(-a_1\gamma_{XY}(k-1) + b_0\gamma_X(k)),$$

and, as above,

$$\begin{aligned}
\gamma_{XY}(k-1) &= \mathbf{E}(X(0)Y(k-1)) \\
&= \mathbf{E}(X(0)(-a_1Y(k-2) + b_0X(k-1))) \\
&= (-a_1)\mathbf{E}(X(0)Y(k-2)) \\
&= (-a_1)\gamma_{XY}(k-2) \\
&= \cdots = (-a_1)^{k-1}\gamma_{XY}(0) = b(0)(-a_1)^{k-1}.
\end{aligned}$$

Since the autocorrelation sequence must be an even function of variable k, we finally get, for any $k = \ldots, -2, -1, 0, 1, 2, \ldots$,

$$\gamma_Y(k) = \frac{b_0^2}{1-a_1^2}(-a_1)^{|k|},$$

thus recovering the result from Chapter 4.

6.4 Problems and exercises

In the first three exercises, also try solving the problem by first finding the autocorrelation function of the output to see how hard the problem is in the time domain framework.

6.4.1. The impulse response function of a linear system is $h(t) = 1 - t$ for $0 \le t \le 1$ and 0 elsewhere.

(a) Produce a graph of $h(t)$.

(b) Assume that the input is the standard white noise. Find the autocorrelation function of the output.

(c) Find the power transfer function of the system, its equivalent-noise bandwidth and half-power bandwidth.

(d) Assume that the input has the autocorrelation function $\gamma_X(t) = \frac{3}{1+4t^2}$. Find the power spectrum of the output signal.

(e) Assume that the input has the autocorrelation function $\gamma_X(t) = \exp(-4|t|)$. Find the power spectrum of the output signal.

(f) Assume that the input has the autocorrelation function $\gamma_X(t) = 1 - |t|$ for $|t| < 1$ and 0 elsewhere. Find the power spectrum of the output signal.

6.4.2. The impulse response function of a linear system is $h(t) = e^{-2t}$ for $0 \le t \le 2$ and 0 elsewhere.

(a) Produce a graph of $h(t)$.

(b) Assume that the input is the standard white noise. Find the autocorrelation function of the output.

(c) Find the power transfer function of the system, its equivalent-noise bandwidth and half-power bandwidth.

(d) Assume that the input has the autocorrelation function $\gamma_X(t) = \frac{3}{1+4t^2}$. Find the power spectrum of the output signal.

(e) Assume that the input has the autocorrelation function $\gamma_X(t) = \exp(-4|t|)$. Find the power spectrum of the output signal.

(f) Assume that the input has the autocorrelation function $\gamma_X(t) = 1 - |t|$ for $|t| < 1$ and 0 elsewhere. Find the power spectrum of the output signal.

6.4.3. The impulse response function of a linear system is $h(t) = e^{-0.05t}$ for $t \ge 10$ and 0 elsewhere.

(a) Produce a graph of $h(t)$.

(b) Assume that the input is the standard white noise. Find the autocorrelation function of the output.

(c) Find the power transfer function of the system, its equivalent-noise bandwidth and half-power bandwidth.

(d) Assume that the input has the autocorrelation function $\gamma_X(t) = \frac{3}{1+4t^2}$. Find the power spectrum of the output signal.

(e) Assume that the input has the autocorrelation function $\gamma_X(t) = \exp(-4|t|)$. Find the power spectrum of the output signal.

(f) Assume that the input has the autocorrelation function $\gamma_X(t) = 1 - |t|$ for $|t| < 1$ and 0 elsewhere. Find the power spectrum of the output signal.

6.4.4. *Cross-correlation* ρ_{XY} and *cross-covariance* γ_{XY} for random signals $X(t)$ and $Y(t)$ are defined, respectively, as follows:

$$\rho_{XY}(t,s) = \mathbf{E}(X(t)Y(s)),$$
$$\gamma_{XY}(t,s) = \mathbf{E}((X(t) - \mu_X(t))(Y(s) - \mu_Y(s)).$$

Random signals $X(t)$ and $Y(t)$ are said to be *jointly stationary* if they are stationary and

$$\rho_{XY}(t,s) = \rho_{XY}(t-s,0).$$

Consider random signals

$$X(t) = a\cos(2\pi(f_0t + \Theta)), \qquad Y(t) = b\sin(2\pi(f_0t + \Theta)),$$

where a and b are nonrandom constants and Θ is uniformly distributed on $[0,1]$. Find the cross-correlation function for X and Y. Are these signals jointly stationary?

6.4.5. Consider the circuit shown in Figure 6.4.1

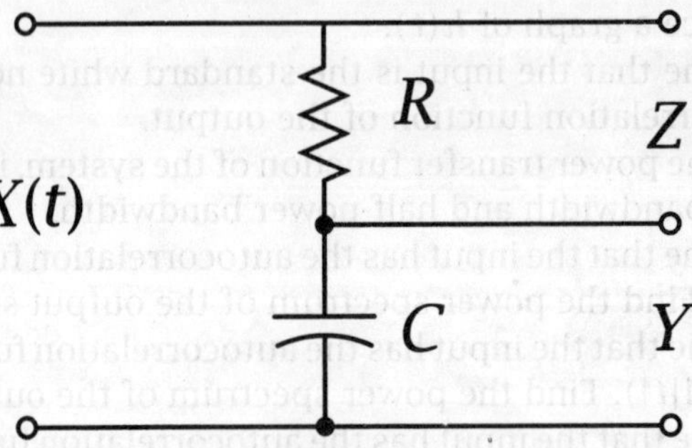

Fig. 6.4.1.

Assume that the input is the standard white noise.

(a) Find the power spectra $S_Y(f)$ and $S_Z(f)$ of the outputs $Y(t)$ and $Z(t)$.

(b) Find the cross-correlation

$$\gamma_{YZ}(\tau) = \mathrm{E}(Z(t)Y(t+\tau))$$

between those two outputs.

6.4.6. Find the output autocorrelation sequence for the discrete-time system representing a stochastic difference equation described in Example 6.3.2. Use the Fourier series expansion of formula (6.3.13).

6.4.7. Consider the circuit shown in Figure 6.4.2.

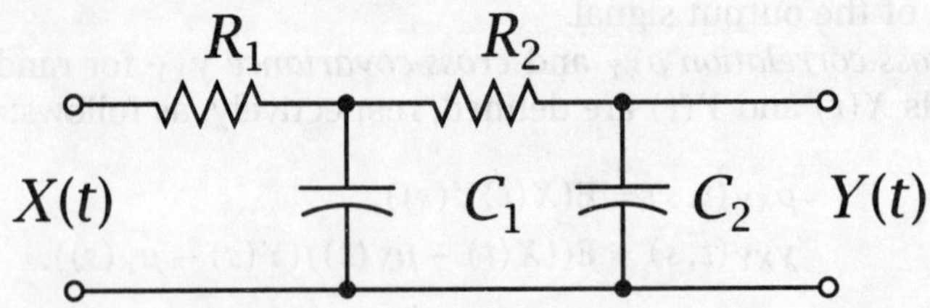

Fig. 6.4.2.

Assume that the input is the standard white noise. Find the power spectrum $S_Y(f)$ and the autocorrelation function $\gamma_Y(\tau)$ of the output $Y(t)$.

6.4.8. Find the half-power and equivalent-noise bandwidth of the system shown in Figure 6.4.2.

7

Optimization of Signal-to-Noise Ratio in Linear Systems

Useful, deterministic signals passing through various transmission devices often acquire extraneous random components due to, say, thermal noise in conducting materials, radio clutter or aurora borealis magnetic field fluctuations in the atmosphere, or deliberate jamming in warfare. If there exists some prior information about the nature of the original useful signal and the contaminating random noise it is possible to devise algorithms to improve the relative power of the useful compenent of the signal or, in other words, to increase the *signal-to-noise ratio* of the signal, by passing it through a filter designed for the purpose. In this short chapter, we give a few examples of such designs just to show how the previously introduced techniques of analysis of random signals can be applied in this context.

7.1 Fixed filter structure, known input signal

The general problem of optimization (maximization) of the signal-to-noise ratio in a linear system schematically pictured here,

$$x(t) + N(t) \longrightarrow \boxed{\text{h(t)}} \longrightarrow y(t) + M(t),$$

can be formulated as follows: Consider a linear filter (system) characterized by its impulse response function $h(t)$ with the input signal $X(t)$ of the form

$$X(t) = x(t) + N(t), \tag{7.1.1}$$

where $x(t)$ is a deterministic "useful" signal and $N(t)$ is a random stationary "noise" signal with zero mean and autocorrelation function $\gamma_N(t)$. Given the linearity of the system, the output signal $Y(t)$ is of the form

$$Y(t) = y(t) + M(t), \tag{7.1.2}$$

where the deterministic "useful" output component is

$$y(t) = \int_{-\infty}^{\infty} x(s)h(t-s)ds, \tag{7.1.3}$$

and the "noise" output is a stationary zero-mean signal with the autocorrelation function

$$\gamma_M(\tau) = \int_{-\infty}^{\infty}\int_{-\infty}^{\infty} \gamma_N(\tau - s + u)h(s)h(u)dsdu.$$

The task is as follows: Given the shape of the input signal, design the structure of the filter which would maximize the signal-to-noise power ratio on the output. More precisely, we need to find an impulse response function $h(t)$ such that, for a given detection time t, the signal-to-noise ratio

$$\frac{S}{\mathcal{N}} = \frac{\mathrm{PW}_y(t)}{\mathrm{E}(\mathrm{PW}_M)} \tag{7.1.4}$$

is maximized over all possible impulse response functions; in brief, we want to find $h(t)$ for which

$$\frac{S}{\mathcal{N}} = \max .$$

Here, $\mathrm{PW}_y(t) = y^2(t)$ is the instantaneous power of the output signal, and $\mathrm{E}(\mathrm{PW}_M) = \gamma_M(0) = \sigma_M^2$ is the mean power of the output noise. Hence the optimization problem is to find $h(t)$, and also the detection time t_0, such that

$$\frac{S}{\mathcal{N}} = \frac{y^2(t_0)}{\gamma_M(0)} = \frac{y^2(t_0)}{\sigma_M^2} = \max . \tag{7.1.5}$$

In the present section we will take a look at a relatively simple situation when the general structure of the filter is essentially fixed and only certain parameters, including the detection time t_0, need to be optimized.

To show the essence of our approach, we will just consider the RC filter with the impulse response function

$$h(t) = be^{-bt} \cdot u(t), \tag{7.1.6}$$

with a single parameter $b = \frac{1}{RC}$ to be determined in addition to the optimal detection time t_0.

Suppose that the "useful" input signal we are trying to detect on the output is a rectangular impulse

$$x(t) = \begin{cases} A & \text{for } 0 \le t \le T; \\ 0 & \text{elsewhere} \end{cases} \tag{7.1.7}$$

and that the input noise is a white noise with the autocovariance $\gamma_N(t) = N_0\delta(t)$.

The deterministic "useful" output signal is

$$\begin{aligned} y(t) &= \int_{-\infty}^{\infty} x(s)h(-(s-t))ds \\ &= \begin{cases} \int_0^t Abe^{-b(t-s)}ds & \text{for } 0 < t < T; \\ \int_0^T Abe^{-b(t-s)}ds & \text{for } t \geq T, \end{cases} \\ &= \begin{cases} A(1-e^{-bt}) & \text{for } 0 < t \leq T; \\ A(1-e^{-bT})e^{-b(t-T)} & \text{for } t \geq T. \end{cases} \end{aligned} \tag{7.1.8}$$

It is pictured in Figure 7.1.1.

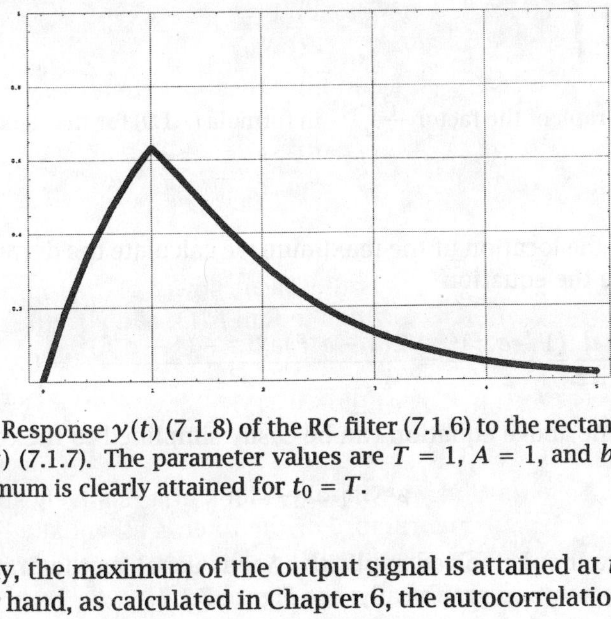

Fig. 7.1.1. Response $y(t)$ (7.1.8) of the RC filter (7.1.6) to the rectangular input signal $x(t)$ (7.1.7). The parameter values are $T = 1$, $A = 1$, and $b = \frac{1}{RC} = 1$. The maximum is clearly attained for $t_0 = T$.

Clearly, the maximum of the output signal is attained at $t_0 = T$. On the other hand, as calculated in Chapter 6, the autocorrelation function of the output noise is

$$\gamma_M(\tau) = N_0\frac{b}{2}e^{-b\tau},$$

so that, at the already optimized detection time $t_0 = T$,

$$\frac{S}{\mathcal{N}} = \frac{y^2(T)}{\gamma_M(0)} = \frac{A^2[1-e^{-bT}]^2}{\frac{bN_0}{2}}.$$

To simplify our calculations we will substitute $z = bT$. Now our final task is to find the maximum of the function

$$\frac{S}{\mathcal{N}}(z) = \frac{2A^2T}{N_0}\cdot\frac{(1-e^{-z})^2}{z} \tag{7.1.9}$$

of one variable z. Function $\frac{S}{\mathcal{N}}(z)$, although simple-looking, is a little tricky and we will start the exploration of its maximum by graphing it; see Figure 7.1.2.

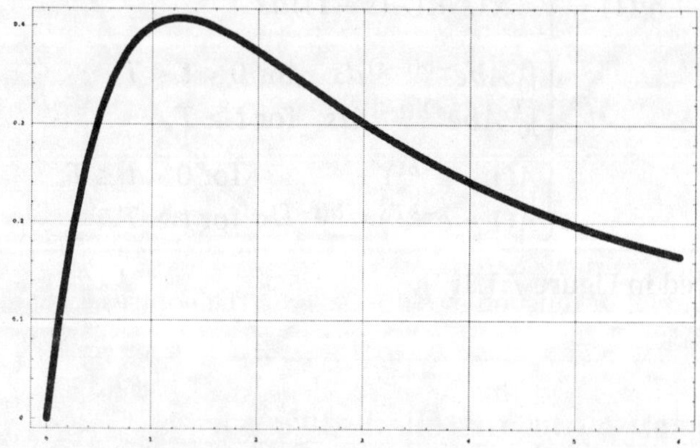

Fig. 7.1.2. Graph of the factor $\frac{(1-e^{-z})^2}{z}$ in formula (7.1.9) for the signal-to-noise ratio $\frac{S}{\mathcal{N}}(z)$.

To find the location of the maximum we calculate the derivative and try to solve the equation

$$\frac{d}{dz}\frac{(1-e^{-z})^2}{z} = \frac{2(1-e^{-z})e^{-z}z-(1-e^{-z})^2}{z^2} = 0.$$

Although the above equation can be easily simplified to the equation

$$e^z - 1 - 2z = 0,$$

the latter cannot be solved explicitly. So, as usual, as the first step we explore the solution graphically; see Figure 7.1.3.

The nontrivial zero is approximately at $z_{\max} = 1.25$, which gives $b_{\max} = \frac{1.25}{T}$ so that the optimal RC constant is

$$RC_{\max} \approx \frac{1}{b_{\max}} = \frac{T}{1.25} = 0.8T. \tag{7.1.10}$$

Evaluated at the optimal values of parameters t_0 and b, the maximum available signal-to-noise ratio is

$$\frac{S}{\mathcal{N}}_{\max} \approx \frac{y^2(T)}{\frac{b_{\max}N_0}{2}} = \frac{2A^2[1-e^{-b_{\max}T}]^2}{b_{\max}N_0} = 0.81 \cdot \frac{A^2T}{N_0}. \tag{7.1.11}$$

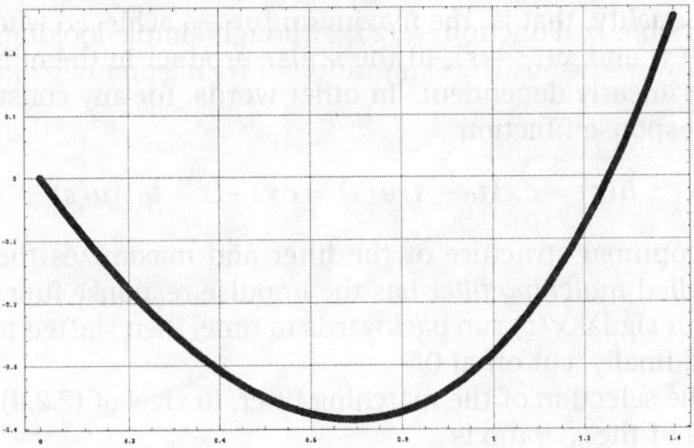

Fig. 7.1.3. A plot of function $e^z - 1 - 2z = 0$. The nontrivial zero is approximately at $z_{\max} = 1.25$.

7.2 Filter structure matched to signal

In this section we will solve a more ambitious problem of designing the structure of the filter to maximize the signal-to-noise ratio on the output rather than just optimizing filter parameters. To be more precise, the task at hand is to find an impulse response function $h(t)$ and the detection time t_0 such that

$$\frac{S}{\mathcal{N}} = \frac{y^2(t_0)}{\sigma_M^2} = \max \tag{7.2.1}$$

for a given deterministic (nonrandom) input signal $x(t)$ transmitted in the presence of the white noise input $N(t)$ with autocorrelation function $\gamma_N(t) = N_0\delta(t)$, where, as before, $x(t) = 0$ for $t \leq 0$ and

$$y(t) = \int_0^\infty x(t-s)h(s)ds. \tag{7.2.2}$$

For the output noise,

$$\sigma_M^2 = \gamma_M(0) = \int_0^\infty \left(\int_0^\infty \delta(u-s)h(u)du\right)h(s)ds = N_0\int_0^\infty h^2(s)ds. \tag{7.2.3}$$

In this situation,

$$\frac{S}{\mathcal{N}} = \frac{y^2(t_0)}{\sigma_M^2} = \frac{(\int_0^\infty x(t_0-s)h(s)ds)^2}{N_0\int_0^\infty h^2(s)ds}. \tag{7.2.4}$$

In view of the Cauchy-Schwartz inequality,

$$\frac{S}{\mathcal{N}} \leq \frac{\int_0^\infty x^2(t_0-s)ds \cdot \int_0^\infty h^2(s)ds}{N_0\int_0^\infty h^2(s)ds} = \frac{1}{N_0}\int_0^\infty x^2(t_0-s)ds, \tag{7.2.5}$$

with the equality, that is, the maximum for $\frac{S}{\mathcal{N}}$, achieved when the two factors, $h(s)$ and $x(t_0 - s)$, in the scalar product in the numerator of (7.2.4) are linearly dependent. In other words, for any constant c, the impulse response function

$$h(s) = cx(t_0 - s)u(s) = cx(-(s - t_0))u(s) \tag{7.2.6}$$

gives the optimal structure of the filter and maximizes the $\frac{S}{\mathcal{N}}$ ratio. This so-called *matching filter* has the impulse response function equal to the input signal $x(t)$ run backwards in time, then shifted to the right by t_0, and, finally, cut off at 0.

With the selection of the matching filter, in view of (7.2.4), the maximal value of the $\frac{S}{\mathcal{N}}$ ratio is

$$\frac{S}{\mathcal{N}}_{\max} = \frac{(\int_0^\infty x(t_0 - s)cx(t_0 - s)u(s)ds)^2}{N_0 \int_0^\infty (cx(t_0 - s)u(s))^2 ds} = \frac{\int_0^\infty x^2(t_0 - s)ds}{N_0}. \tag{7.2.7}$$

Example 7.2.1 (matching filter for a rectangular input signal). Consider a rectangular input signal of the form

$$x(t) = \begin{cases} A & \text{for } 0 < t < T; \\ 0 & \text{elsewhere,} \end{cases}$$

transmitted in the presence of an additive white noise with autocorrelation function $\gamma_N(t) = N_0\delta(t)$. According to formula (7.2.6), its matching filter at detection time t_0 is

$$h(t) = \begin{cases} A & \text{for } 0 < t < t_0; \\ 0 & \text{elsewhere} \end{cases}$$

if $0 \le t_0 \le T$ and

$$h(t) = \begin{cases} A & \text{for } t_0 - T < t < t_0; \\ 0 & \text{elsewhere} \end{cases}$$

if $t_0 > T$. So the $\frac{S}{\mathcal{N}}_{\max}$, as a function of the detection time t_0, is

$$\frac{S}{\mathcal{N}}_{\max}(t_0) = \begin{cases} \frac{A^2 t_0}{N_0} & \text{for } 0 < t_0 < T; \\ \frac{A^2 T}{N_0} & \text{for } t_0 > T. \end{cases}$$

Clearly, the earliest detection time t_0 to maximize $\frac{S}{\mathcal{N}}_{\max}(t_0)$ is $t_0 = T$ (see Figure 7.2.1).

At the optimal detection time $t_0 = T$, or any later detection time,

$$\frac{S}{\mathcal{N}}_{\max} = \frac{A^2 T}{N_0}. \tag{7.2.8}$$

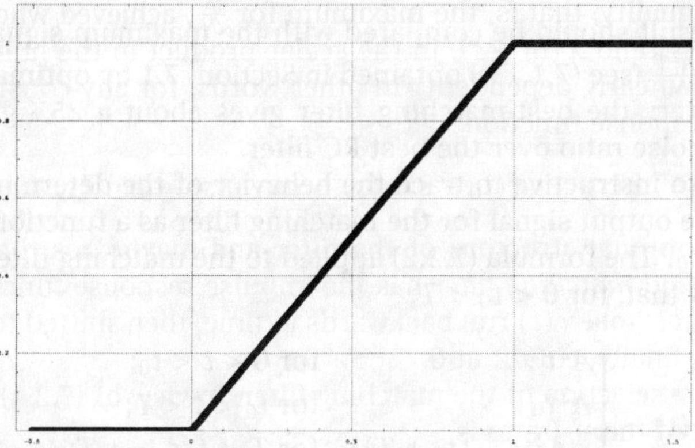

Fig. 7.2.1. The dependence of the optimal signal-to-noise ratio on the detection time t_0 for the matching filter from Example 7.2.1. The input signal is the sum of a rectangular signal of amplitude $A = 1$, duration $T = 1$, and the white noise with autocorrelation function $\gamma_N(t) = \delta(t)$.

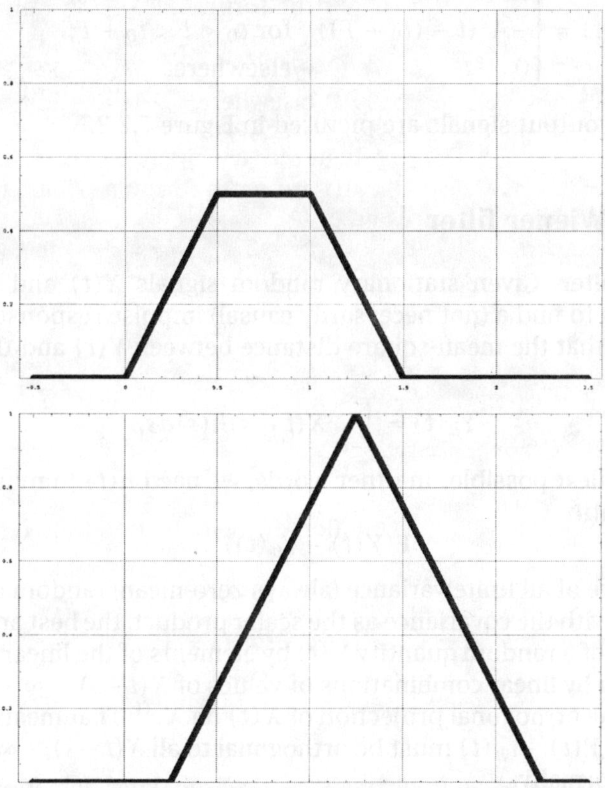

Fig. 7.2.2. The response $y(t)$ of the matching filter for the rectangular input signal with amplitude $A = 1$ and duration $T = 1$ (see Example 7.2.1). *Top*: For detection time $t_0 = 0.25 < T = 1$. *Bottom*: For detection time $t_0 = 1.25 > T = 1$.

This result should be compared with the maximum signal-to-noise ratio $0.81\frac{A^2T}{N_0}$ (see (7.1.11)) obtained in Section 7.1 by optimally tuning the RC filter: the best-matching filter gives about a 25% gain in the signal-to-noise ratio over the best RC filter.

It is also instructive to trace the behavior of the deterministic part $y(t)$ of the output signal for the matching filter as a function of detection time t_0. The formula (7.2.2) applied to the matching filter immediately gives that, for $0 < t_0 < T$,

$$y(t) = \begin{cases} A^2 t & \text{for } 0 < t < t_0; \\ A^2 t_0 & \text{for } t_0 < t < T; \\ -A^2(t - (t_0 + T)) & \text{for } T < t < t_0 + T; \\ 0 & \text{elsewhere} \end{cases} \tag{7.2.9}$$

and, for $t_0 \geq T$,

$$y(t) = \begin{cases} A^2(t - (t_0 - T)) & \text{for } t_0 - T < t < t_0; \\ -A^2(t - (t_0 + T)) & \text{for } t_0 < t < t_0 + T; \\ 0 & \text{elsewhere.} \end{cases} \tag{7.2.10}$$

These two output signals are pictured in Figure 7.2.2.

7.3 The Wiener filter

Acausal filter. Given stationary random signals $X(t)$ and $Y(t)$, the problem is to find a (not necessarily causal) impulse response function $h(t)$ such that the mean-square distance between $Y(t)$ and the output signal,

$$Y_h(t) = \int_{-\infty}^{\infty} X(t-s)h(s)ds,$$

is the smallest possible. In other words, we need $h(t)$ to minimize the error quantity

$$\mathbf{E}(Y(t) - Y_h(t))^2.$$

In the space of all finite variance (always zero-mean) random quantities equipped with the covariance as the scalar product, the best approximation $Y_h(t)$ of a random quantity $Y(t)$ by elements of the linear subspace $\mathcal{X}$ spanned by linear combinations of values of $X(t-s)$, $-\infty < s < \infty$, is given by the orthogonal projection of $X(t)$ on $\mathcal{X}$.[31] That means that the difference $Y(t) - Y_h(t)$ must be orthogonal to all $X(t-s)$, $-\infty < s < \infty$, or, more formally,

[31] This argument is analagous to the one encountered in Chapter 2, when we discussed the best approximation in power (for a definition, see Section 2.2) of deterministic periodic signals by their Fourier series.

$$\begin{aligned}
&\mathbf{E}((Y(t) - Y_h(t)) \cdot X(t-s)) \\
&= \mathbf{E}(Y(t) \cdot X(t-s)) - \mathbf{E}\left(\int_{-\infty}^{\infty} X(t-u)h(u)du \cdot X(t-s)\right) \\
&= \gamma_{YX}(s) - \int_{-\infty}^{\infty} \gamma_X(s-u)h(u)du = 0,
\end{aligned}$$

for all s, $-\infty < s < \infty$. Hence the optimal $h(t)$ can be found by solving, for each s, the integral equation

$$\gamma_{YX}(s) = \int_{-\infty}^{\infty} \gamma_X(s-u)h(u)du, \tag{7.3.1}$$

which involves only the autocorrelation function $\gamma_X(s)$ and the cross-correlation function $\gamma_{YX}(s)$. The solution is found readily in the frequency domain. Remembering that the Fourier transform of a convolution is the product of Fourier transforms, and denoting by $H(f)$ the transfer function (the Fourier transform of the impulse response function) of the optimal $h(t)$, (7.3.1) can be rewritten in the form

$$S_{YX}(f) = S_X(f) \cdot H(f),$$

which immediately gives the explicit formula for the transfer function of the optimal filter:

$$H(f) = \frac{S_{YX}(f)}{S_X(f)}. \tag{7.3.2}$$

The minimal error can then also be calculated explicitly:

$$\mathbf{E}(Y(t) - Y_h(t))^2 = \gamma_Y(0) - \int_{-\infty}^{\infty} \gamma_{YX}(s)h(s)ds, \tag{7.3.3}$$

or, in terms of the optimal transfer function, using the Parseval formula for the last integral, we have

$$\mathbf{E}(Y(t) - Y_h(t))^2 = \int_{-\infty}^{\infty} (S_Y(f) - S_{YX}^*(f)H(f))df. \tag{7.3.4}$$

Example 7.3.1. Assume that signal $X(t)$ is the sum of a "useful" signal $Y(t)$ and noise $N(t)$, that is, $X(t) = Y(t) + N(t)$, where $Y(t)$ has the power spectrum

$$S_Y(f) = \frac{1}{1+f^2},$$

and is uncorrelated with the white noise $N(t)$, which is assumed to have the power spectrum $S_N(f) \equiv 1$. Then

$$S_{YX}(f) = S_Y(f) = \frac{1}{1+f^2} \quad \text{and} \quad S_X(f) = S_Y(f) + S_N(f) = \frac{2+f^2}{1+f^2}.$$

The transfer function of the optimal filter is then

$$H(f) = \frac{S_{YX}(f)}{S_X(f)} = \frac{1}{2+f^2},$$

with the corresponding impulse response function

$$h(t) = \frac{1}{2\sqrt{2}} e^{-\sqrt{2}|t|},$$

and the error

$$\mathbf{E}(Y(t) - Y_h(t))^2 = \int_{-\infty}^{\infty} \left(\frac{1}{1+f^2} - \frac{1}{1+f^2} \cdot \frac{1}{2+f^2} \right) df$$
$$= \int_{-\infty}^{\infty} \frac{1}{2+f^2} df = \frac{\pi}{\sqrt{2}}.$$

Causal filter. For given stationary random signals $X(t)$ and $Y(t)$, the construction of the optimal *causal filter* requires finding a causal impulse response function $h(t) = 0$, for $t \le 0$, such that the error

$$\mathbf{E}\left(Y(t) - \int_0^{\infty} X(t-s)h(s)ds \right)^2$$

is minimal. In other words, we are trying to find the best mean-square approximation to $Y(t)$ by (continuous) linear combinations of the *past* values of $X(t)$. Using the same orthogonality argument we applied for the acausal optimal filter, we obtain another integral equation for the optimal $h(t)$:

$$\gamma_{YX}(s) = \int_0^{\infty} \gamma_X(s-u)h(u)du,$$

this time valid only for all $s > 0$. This equation is traditionally called the *Wiener-Hopf equation*. It is clear that to solve the above equation via an integral transform method we have to replace the Fourier transform used in the acausal case by the Laplace transform. However, the details here are more involved, and for the solution, we refer the reader to the literature of the subject.[32]

7.4 Problems and exercises

7.4.1. The triangular signal $x(t) = 0.01t$ for $0 < t < 0.01$ and 0 elsewhere is combined with white noise having a flat power spectrum of $2\frac{V^2}{\text{Hz}}$. Find the value of the RC-constant such that the signal-to-noise ratio at the output of the RC filter is maximal at $t = 0.01$ second.

[32] N. Wiener's original *Extrapolation, Interpolation, and Smoothing of Stationary Time Series*, MIT Press and Wiley, New York, 1950, is still very readable, but also see Chapter 10 of A. Papoulis, *Signal Analysis*, McGraw-Hill, New York, 1977.

7.4.2. A signal of the form $x(t) = 5e^{-(t+2)}u(t)$ is to be detected in the presence of white noise with a flat power spectrum of $0.25\frac{V^2}{\text{Hz}}$ using a matched filter.

(a) For $t_0 = 2$ find the value of the impulse response of the matched filter at $t = 0, 2, 4$.

(b) Find the maximum output signal-to-noise ratio that can be achieved if $t_0 = \infty$.

(c) Find the detection time t_0 that should be used to achieve an output signal-to-noise ratio that is equal to 95% of the maximum signal-to-noise ratio discovered in part (b).

(d) The signal $x(t) = 5e^{-(t+2)}u(t)$ is combined with white noise having a power spectrum of $2\frac{V^2}{\text{Hz}}$. Find the value of RC such that the signal/noise at the output of the RC filter is maximal at $t = 0.01$ second.

7.4.3. Repeat construction of the optimal filter from Example 7.3.1 in the case when the useful signal $Y(t)$ has a more general power spectrum

$$S_Y(f) = \frac{a}{b^2 + f^2},$$

and the uncorrelated white noise $N(t)$ has arbitrary power spectrum $S_N(f) \equiv \mathcal{N}$. Discuss the properties of this filter when the noise power is much bigger than the power of the useful signal, that is, when $\mathcal{N} \gg S_Y(f)$. Construct the optimal acausal filters for other selected spectra of $Y(t)$ and $N(t)$.

8

Gaussian Signals, Correlation Matrices, and Sample Path Properties

In general, determination of the shape of the sample paths of a random signal $X(t)$ requires knowledge of n-point probabilities

$$\mathbf{P}(a_1 < X(t_1) < b_1, \ldots, a_n < X(t_n) < b_n)$$

for an arbitrary n, and arbitrary windows $a_1 < b_1, \ldots, a_n < b_n$. But usually this information cannot be recovered if the only signal characteristic known is the autocorrelation function. The latter depends on the two-point distributions but does not uniquely determine them. However, in the case of Gaussian signals, the autocorrelations determine not only the two-point probability distributions but also all the n-point probability distributions, so that complete information is available within the second-order theory. For example, this means that you only have to estimate means and covariances to make the model. Also, in the Gaussian universe, weak stationarity implies strict stationarity as defined in Chapter 4. For the sake of simplicity all signals in this chapter are assumed to be real-valued. The chapter ends with a more subtle analysis of sample path properties of stationary signals such as continuity and differentiability; in the Gaussian case the information is particularly complete.

Of course, faced with real-world data the proposition that they are distributed according to a Gaussian distribution must be tested rigorously. Many such tests have been developed by the statisticians.[33] In other cases, one can make an argument in favor of such a hypothesis based on the central limit theorem (3.5.5)–(3.5.6).

[33] See, e.g., M. Denker and W. A. Woyczyński's book mentioned in previous chapters.

8.1 Linear transformations of random vectors

In Chapter 3, we have calculated probability distributions of transformed random quantities. Repeating that procedure in the case of a linear transformation of the 1D random quantity X given by the formula

$$Y = aX, \quad a > 0, \tag{8.1.1}$$

we can obtain the cumulative distribution function (c.d.f.) $F_Y(y)$ of the random quantity Y in terms of the c.d.f. $F_X(x)$ of the random quantity X as follows:

$$F_Y(y) = P(Y \le y) = P(aX \le y) = P\left(X \le \frac{y}{a}\right) = F_X\left(\frac{y}{a}\right). \tag{8.1.2}$$

To obtain an analogous formula for the probability density functions (p.d.f.s), it suffices to differentiate both sides of (8.1.2) to see that

$$f_Y(y) = \frac{d}{dy}F_Y(y) = \frac{1}{a}f_X\left(\frac{u}{a}\right). \tag{8.1.3}$$

Example 8.1.1. Consider a standard Gaussian random quantity $X \sim N(0,1)$ with the p.d.f.

$$f_X(x) = \frac{1}{\sqrt{2\pi}}\exp\left(\frac{-x^2}{2}\right). \tag{8.1.4}$$

Then the random quantity $Y = aX$, $a > 0$, has the p.d.f.

$$f_Y(y) = \frac{1}{\sqrt{2\pi}a}\exp\left(-\frac{x^2}{2a^2}\right). \tag{8.1.5}$$

Obviously, the expectation $\mathbf{E}Y = \mathbf{E}(aX) = a\mathbf{E}X = 0$ and the variance of Y is

$$\sigma_Y^2 = \mathbf{E}(aX)^2 = a^2\mathbf{E}X^2 = a^2. \tag{8.1.6}$$

If we conduct the same argument for $a < 0$, the p.d.f. of $Y = aX$ will be

$$f_Y(y) = \frac{1}{\sqrt{2\pi}(-a)}\exp\left(-\frac{x^2}{2a^2}\right). \tag{8.1.7}$$

Thus formulas (8.1.6) and (8.1.7) can be unified in a single statement: If $X \sim N(0,1)$, then for any $a \neq 0$, random quantity $Y = aX$ has p.d.f.

$$f_Y(y) = \frac{1}{\sqrt{2\pi}|a|}\exp\left(-\frac{x^2}{2a^2}\right). \tag{8.1.8}$$

Using the above elementary reasoning as a model we will now derive the formula for a d-dimensional p.d.f.

$$f_{\vec{Y}}(\vec{y}) = f_{\vec{Y}}(y_1, \ldots, y_d)$$

of a random vector

$$\vec{Y} = \begin{pmatrix} Y_1 \\ \vdots \\ Y_d \end{pmatrix}$$

obtained by a nondegenerate (invertible) linear transformation

$$\vec{Y} = \mathbf{A}\vec{X} \tag{8.1.9}$$

consisting of multiplication of the random vector

$$\vec{X} = \begin{pmatrix} X_1 \\ \vdots \\ X_d \end{pmatrix}$$

with a known p.d.f.

$$f_{\vec{X}}(\vec{x}) = f_{\vec{X}}(x_1, \dots, x_d)$$

by a fixed nondegenerate nonrandom matrix

$$\mathbf{A} = \begin{pmatrix} a_{11}, \cdots, a_{1d} \\ \cdots \\ a_{d1}, \cdots, a_{dd} \end{pmatrix}.$$

In other words, we assume that $\det(A) \neq 0$, or, equivalently, that the rows of the matrix A form a linearly independent system of vectors.

In terms of its coordinates the result of the linear transformation (8.1.9) can be written in the explicit form

$$\vec{Y} = \begin{pmatrix} a_{11}X_1 + a_{12}X_2 + \cdots + a_{1d}X_d \\ a_{21}X_1 + a_{22}X_2 + \cdots + a_{2d}X_d \\ \cdots \quad \cdots \quad \cdots \quad \cdots \\ a_{d1}X_1 + a_{d2}X_2 + \cdots + a_{dd}X_d \end{pmatrix}.$$

To calculate the probability distribution of $\vec{Y}$ following the 1D method, we must use the assumption that the matrix $\mathbf{A}$ is invertible, an analogue of the assumption $a \neq 0$ in the 1D case. Then, for a domain D in the d-dimensional space $\mathbf{R}^d$,

$$\mathbf{P}(\vec{Y} \in D) = \mathbf{P}(\mathbf{A}\vec{X} \in D) = \mathbf{P}(\vec{X} \in \mathbf{A}^{-1}D). \tag{8.1.10}$$

This identity can be rewritten in terms of p.d.f.s of $\vec{Y}$ and $\vec{X}$ as follows:

$$\int_D f_{\vec{Y}}(\vec{y})dy_1 \cdot \cdots \cdot dy_d = \int_{\mathbf{A}^{-1}D} f_{\vec{X}}(\vec{x})dx_1 \cdot \cdots \cdot dx_d.$$

Making a substitution $\vec{x} = \mathbf{A}^{-1}\vec{z}$ in the second integral, in view of the d-dimensional change of variables formula, we get that

$$\int_D f_{\vec{Y}}(\vec{y})dy_1\cdot\dots\cdot dy_d = \int_D f_{\vec{X}}(\mathrm{A}^{-1}\vec{z})\cdot|\det(\mathrm{A}^{-1})|dz_1\cdot\dots\cdot dz_d,$$

where $\det(\mathrm{A}^{-1})$ is just the Jacobian of the substitution $\vec{x} = \mathrm{A}^{-1}\vec{z}$. Remembering that the determinant of the inverse matrix A^{-1} is the reciprocal of the determinant of the matrix A, we get the identity

$$\int_D f_{\vec{Y}}(\vec{y})dy_1\cdot\dots\cdot dy_d = \int_D \frac{f_{\vec{X}}(\mathrm{A}^{-1}\vec{z})}{|\det(\mathrm{A})|}dz_1\cdot\dots\cdot dz_d.$$

Since this identity holds true for any domain D, the integrands on both sides must be equal, which gives the final formula for the p.d.f. of $\vec{Y}$:

$$f_{\vec{Y}}(\vec{y}) = \frac{f_{\vec{X}}(\mathrm{A}^{-1}\vec{y})}{|\det(\mathrm{A})|} \quad \text{if } \det(\mathrm{A}) \neq 0. \tag{8.1.11}$$

The 1D formula (8.1.3) is, obviously, the special case of the above general result.

8.2 Gaussian random vectors

As in the one-dimensional case, all nondegenerate zero-mean d-dimensional Gaussian random vectors can be obtained as nondegenerate linear transformations of a standard dD Gaussian random vector

$$\vec{X} = \begin{pmatrix} X_1 \\ \vdots \\ X_d \end{pmatrix}$$

in which the coordinates $X_1,\dots,X_d$, are independent $N(0,1)$ random quantities. Because of their independence, the d-dimensional p.d.f. of $\vec{X}$ is the product of 1D $N(0,1)$ p.d.f.s and is thus of the product form

$$\begin{aligned} f_{\vec{X}}(\vec{x}) &= \frac{e^{-x_1^2/2}}{\sqrt{2\pi}}\cdot\dots\cdot\frac{e^{-x_d^2/2}}{\sqrt{2\pi}} \\ &= \frac{1}{(2\pi)^{d/2}}e^{-\frac{1}{2}(x_1^2+\dots+x_d^2)} = \frac{1}{(2\pi)^{d/2}}e^{-\frac{1}{2}\vec{x}^T\vec{x}}, \end{aligned} \tag{8.2.1}$$

where the superscript T denotes the transpose of a matrix. Indeed,

$$\vec{x}^T\vec{x} = (x_1,\dots,x_d)\cdot\begin{pmatrix} x_1 \\ \vdots \\ x_d \end{pmatrix} = x_1^2+\dots+x_d^2.$$

It is the latter form in (8.2.1) that will be useful now in applying formula (8.1.11). Indeed, substituting the last expression for $f_{\vec{X}}(\vec{x})$ in (8.2.1) into (8.1.11), one immediately gets, remembering that $(\mathbf{MN})^T = \mathbf{N}^T\mathbf{M}^T$, $(\mathbf{MN})^{-1} = \mathbf{N}^{-1}\mathbf{M}^{-1}$, and $(\mathbf{M}^T)^{-1} = (\mathbf{M}^{-1})^T$,

$$\begin{aligned} f_{\vec{Y}}(\vec{y}) &= \frac{1}{(2\pi)^{d/2}|\det(A)|} e^{-\frac{1}{2}(\mathbf{A}^{-1}\vec{y})^T\cdot(\mathbf{A}^{-1}\vec{y})} \\ &= \frac{1}{(2\pi)^{d/2}|\det(A)|} e^{-\frac{1}{2}\vec{y}^T(\mathbf{A}\mathbf{A}^T)^{-1}\vec{y}}. \end{aligned} \tag{8.2.2}$$

Thus formula (8.2.2) gives the general form of the d-dimensional zero-mean Gaussian p.d.f., and just as we identified the parameter a^2 in the 1D case (8.1.5)–(8.1.6) as the variance of the random quantity Y, we can identify entries of the matrix

$$\mathbf{\Gamma} = \mathbf{A}\mathbf{A}^T \tag{8.2.3}$$

appearing in the exponent in (8.2.2) as statistically significant parameters of the random vector $\vec{Y}$.

To see what they are, let us first calculate the entries γ_{ij}, $i, j = 1, 2, \ldots, d$, of matrix $\mathbf{\Gamma}$:

$$\gamma_{ij} = a_{i1}a_{j1} + a_{i2}a_{j2} + \cdots + a_{id}a_{jd}. \tag{8.2.4}$$

On the other hand, correlations (really, covariances, since we are working with zero-mean vectors) of different components of random vector $\vec{Y}$ are

$$\begin{aligned} \mathbf{E}(Y_iY_j) &= \mathbf{E}((a_{i1}X_1 + \cdots + a_{id}X_d)\cdot(a_{j1}X_1 + \cdots + a_{jd}X_d)) \\ &= a_{i1}a_{j1} + a_{i2}a_{j2} + \cdots + a_{id}a_{jd} \end{aligned} \tag{8.2.5}$$

because $\mathbf{E}X_iX_j = 1$ if $i = j$ and $= 0$ if $i \neq j$.

Therefore, it turns out that

$$\mathbf{\Gamma} = (\gamma_{ij}) = (\mathbf{E}Y_iY_j), \tag{8.2.6}$$

and matrix $\Gamma = (\gamma_{ij})$ is simply the correlation matrix of the general zero-mean Gaussian random vector $\vec{Y}$. Thus, since

$$\det(\mathbf{\Gamma}) = \det(\mathbf{A}\mathbf{A}^T) = \det(\mathbf{A})\cdot\det(\mathbf{A}^T) = (\det(\mathbf{A}))^2,$$

we finally get that the p.d.f. of $\vec{Y}$ can be written in the form

$$f_{\vec{Y}}(\vec{y}) = \frac{1}{(2\pi)^{d/2}|\det(\mathbf{\Gamma})|^{1/2}} e^{-\frac{1}{2}\vec{y}^T\mathbf{\Gamma}^{-1}\vec{y}}, \tag{8.2.7}$$

where $\mathbf{\Gamma}$ is the correlation matrix of $\vec{Y}$ satisfying the nondegeneracy condition $\det(\mathbf{\Gamma}) \neq 0$.

Remark 8.2.1 (*Gaussian random vectors with nonzero mean*). Of course, to get the p.d.f. of a general Gaussian random vector with nonzero expectation

$$\mathbf{E}\vec{Y} = \vec{\mu} = (\mu_1, \dots, \mu_d)^T,$$

it suffices to shift the p.d.f. (8.2.7) by $\vec{\mu}$ to obtain that

$$f_{\vec{Y}}(\vec{y}) = \frac{1}{(2\pi)^{d/2}|\det(\Sigma)|^{1/2}} e^{-\frac{1}{2}(\vec{y}-\vec{\mu})^T\Sigma^{-1}(\vec{y}-\vec{\mu})}, \tag{8.2.8}$$

where

$$\Sigma = (\sigma_{ij}) = (\mathbf{E}(Y_i - \mu_i)(Y_j - \mu_j)) \tag{8.2.9}$$

is the *covariance matrix* of $\vec{Y}$. A Gaussian random vector with joint p.d.f. given by formulas (8.2.8)-(8.2.9) is often called a normal $N(\vec{\mu}, \Sigma)$ random vector.

Example 8.2.1 (2D zero-mean Gaussian random vectors). Let us carry out the above calculation explicitly in the special case of dimension $d = 2$. Then the correlation matrix is

$$\Gamma = \begin{pmatrix} \mathbf{E}Y_1Y_1 & \mathbf{E}Y_1Y_2 \\ \mathbf{E}Y_2Y_1 & \mathbf{E}Y_2Y_2 \end{pmatrix} = \begin{pmatrix} \sigma_1^2 & \sigma_1\sigma_2\rho \\ \sigma_1\sigma_2\rho & \sigma_2^2 \end{pmatrix},$$

where the variances of the coordinate vectors are

$$\sigma_1^2 = \mathbf{E}Y_1^2, \quad \sigma_2^2 = \mathbf{E}Y_2^2,$$

and the correlation coefficient of the two components is

$$\rho = \frac{\mathbf{E}Y_1Y_2}{\sigma_1\sigma_2}.$$

The determinant of the correlation matrix is

$$\det(\Gamma) = \sigma_1^2\sigma_2^2(1 - \rho^2),$$

and its inverse is

$$\Gamma^{-1} = \frac{1}{\sigma_1^2\sigma_2^2(1-\rho^2)} \begin{pmatrix} \sigma_2^2 & -\sigma_1\sigma_2\rho \\ -\sigma_1\sigma_2\rho & \sigma_1^2 \end{pmatrix}.$$

Hence the p.d.f. of a general zero-mean Gaussian random vector is of the form

$$f_{\vec{Y}}(y_1, y_2) = \frac{1}{(2\pi)^{2/2}\sigma_1\sigma_2\sqrt{1-\rho^2}} \times \exp\left[-\frac{1}{2}(y_1, y_2) \frac{\begin{pmatrix} \sigma_2^2 & -\sigma_1\sigma_2\rho \\ -\sigma_1\sigma_2\rho & \sigma_1^2 \end{pmatrix}}{\sigma_1^2\sigma_2^2(1-\rho^2)} \begin{pmatrix} y_1 \\ y_2 \end{pmatrix} \right],$$

which, after performing prescribed matrix algebra, leads to the final expression

$$f_{\mathbf{Y}}(y_1, y_2) = \frac{1}{2\pi\sigma_1\sigma_2\sqrt{1-\rho^2}} \times \exp\left[-\frac{1}{2(1-\rho^2)}\left(\frac{y_1^2}{\sigma_1^2} - 2\rho\frac{y_1 y_2}{\sigma_1\sigma_2} + \frac{y_2^2}{\sigma_2^2}\right)\right]. \quad (8.2.10)$$

8.3 Gaussian stationary signals

By definition, a nondegenerate zero-mean random signal $X(t)$ is Gaussian if, for any positive integer N, and any selection of sampling times $t_1 < t_2 < \cdots < t_N$, the random vector

$$\vec{X}_{(t_1,\ldots,t_N)} = \begin{pmatrix} X(t_1) \\ X(t_2) \\ \vdots \\ X(t_N) \end{pmatrix} \quad (8.3.1)$$

is a Gaussian zero-mean random vector with a nondegenerate correlation matrix. Thus, in view of results of Section 8.2, its N-dimensional joint p.d.f. $f_{(t_1,\ldots,t_N)}(x_1,\ldots,x_N)$ is given by the formula[34]

$$f_{(t_1,\ldots,t_N)}(x_1,\ldots,x_N) = \frac{1}{(2\pi)^{N/2}|\det(\mathbf{\Gamma})|^{1/2}} \cdot e^{-\frac{1}{2}\vec{x}^T\mathbf{\Gamma}^{-1}\vec{x}}, \quad \det(\mathbf{\Gamma}) \neq 0, \quad (8.3.2)$$

where $\mathbf{\Gamma}$ is the $N \times N$ correlation matrix

$$\mathbf{\Gamma} = \mathbf{\Gamma}_{(t_1,\ldots,t_N)} = (\gamma_X(t_i, t_j)) = (\mathbf{E}X(t_i)X(t_j)). \quad (8.3.3)$$

Thus, in view of (8.3.1)–(8.3.2), *the only information needed to completely determine all finite-dimensional joint probability distributions of a zero-mean Gaussian random signal $X(t)$ is the knowledge of its autocorrelation function*

$$\gamma_X(s,t) = \mathbf{E}X(t)X(s).$$

For stationary Gaussian signals the situation is simpler yet as the autocorrelation function $\gamma_X(s,t)$ is just a function of a single variable:

34 Note that for some simple Gaussian stationary signals, like, e.g., $X(t) = X \cdot e^{jt}$, where $X \sim N(0,1)$, one can choose the t_is so that the determinant of the correlation matrix is zero; take, for example, $N = 2$ and $t_1 = \pi$, $t_2 = 2\pi$. Then the joint p.d.f. of the Gaussian random vector $(X(t_1),\ldots,X(t_N))^T$ is not of the form (8.3.2). Such signals are called degenerate.

$$\gamma_X(t,s) = \gamma_X(t-s).$$

Thus the correlation matrix $\mathbf{\Gamma}$ for a stationary random signal $X(t)$ sampled at $t_1, t_2, \dots, t_N$, is of the form

$$\mathbf{\Gamma}_{(t_1,\dots,t_N)} = \begin{pmatrix} \gamma_X(0) & \gamma_X(t_2-t_1) & \gamma_X(t_3-t_1) & \cdots & \gamma_X(t_N-t_1) \\ \dots & \gamma_X(0) & \dots & \dots & \dots \\ \dots & \dots & \dots & \dots & \dots \\ \gamma_X(t_N-t_1) & \gamma_X(t_N-t_2) & \dots & \cdots & \gamma_X(0) \end{pmatrix},$$

and it is obviously invariant under translations, that is, for any t,

$$\mathbf{\Gamma}_{(t_1,\dots,t_N)} = \mathbf{\Gamma}_{(t_1+t,\dots,t_N+t)}, \tag{8.3.4}$$

which, in view of (8.3.2)-(8.3.3), implies that all finite-dimensional p.d.f.s of $X(t)$ are also invariant under translations; that is, for any positive integer N, any sampling times $t_1, \dots, t_N$, and any time shift t,

$$f_{(t_1,\dots,t_N)}(x_1,\dots,x_N) = f_{(t_1+t,\dots,t_N+t)}(x_1,\dots,x_N). \tag{8.3.5}$$

In other words,

a Gaussian weakly stationary signal is strictly stationary

.

In the particular case when the sampling times are uniformly spaced with the intersampling time interval Δt, the correlation matrix $\mathbf{\Gamma}$ of the signal $X(t)$ sampled at times

$$t, \quad t+\Delta t, \quad t+2\Delta t, \quad \dots, \quad t+(N-1)\Delta t,$$

is

$$\mathbf{\Gamma} = \begin{pmatrix} \gamma_X(0) & \gamma_X(\Delta t) & \gamma_X(2\Delta t) & \cdots & \gamma_X((N-1)\Delta t) \\ \gamma_X(\Delta t) & \gamma_X(0) & \gamma_X(\Delta t) & \cdots & \gamma_X((N-2)\Delta t) \\ \dots & \dots & \dots & \dots & \dots \\ \gamma_X((N-1)\Delta t) & \gamma_X((N-2)\Delta t) & \dots & \dots & \gamma_X(0) \end{pmatrix}.$$

Example 8.3.1. Consider a Gaussian signal $X(t)$ with autocorrelation function

$$\gamma_X(t) = e^{-0.3|t|}.$$

We are interested in finding the joint p.d.f. of the signal at times $t_1 = 1$, $t_2 = 2$, and the probability that the signal has values between -0.6 and 1.4 at t_1 and between 0.7 and 2.6 at t_2.

The first step is then to find the correlation matrix

$$\mathbf{\Gamma}_{(1,2)} = \begin{pmatrix} \gamma_X(0) & \gamma_X(1) \\ \gamma_X(1) & \gamma_X(0) \end{pmatrix} = \begin{pmatrix} e^0 & e^{-0.3} \\ e^{-0.3} & e^0 \end{pmatrix} = \begin{pmatrix} 1 & 0.74 \\ 0.74 & 1 \end{pmatrix}.$$

The correlation coefficient of $X(1)$ and $X(2)$ is then

$$\rho = \frac{\gamma_X(1)}{\gamma_X(0)} = 0.74$$

and, in view of Example 8.2.1 (see (8.2.10)), the joint p.d.f. of $X(1)$ and $X(2)$ is of the form

$$\begin{aligned} f_{(1,2)}(x_1, x_2) &= \frac{1}{2\pi\sqrt{1-0.74^2}} \\ &\quad \cdot \exp\left[\frac{-1}{2(1-0.74^2)}(x_1^2 - 2\cdot 0.74x_1x_2 + x_2^2)\right] \\ &= 0.24\cdot\exp[-1.11(x_1^2 - 1.48x_1x_2 + x_2^2)]. \end{aligned}$$

Finally, the desired probability is

$$\begin{aligned} &\mathbf{P}(-0.6 \le X(1) \le 1.4 \text{ and } 0.7 \le X(2) \le 2.6) \\ &= \int_{-0.6}^{1.4}\int_{0.7}^{2.6} 0.24\cdot e^{-1.11(x_1^2-1.48x_1x_2+x_2^2)}\,dx_1dx_2 = 0.17, \end{aligned}$$

where the last integral has been evaluated numerically in *Mathematica* with a two-digit precision.

8.4 Sample path properties of general and Gaussian stationary signals

Mean-square continuity and differentiability. It is clear that the local properties of the autocorrelation function $\gamma_X(\tau)$ of a stationary signal $X(t)$ affect properties of the sample paths of the signal itself in the mean-square sense, that is in terms of the behavior of the expectation of the square of signal's increments, i.e., the variances of the increments.[35] Indeed, with no distributional assumptions on $X(t)$, we have

$$\sigma^2(\tau) = \mathbf{E}(X(t+\tau) - X(t))^2 = 2(\gamma_X(0) - \gamma_X(\tau));$$

the variance of the increment is independent of t. Hence we have the following result:

A stationary signal $X(t)$ is continuous in the mean-square sense, that is, for any $t > 0$,

$$\lim_{\tau\to 0}\mathbf{E}(X(t+\tau) - X(t))^2 = 0,$$

if and only if the autocorrelation function $\gamma_X(\tau)$ is continuous at $\tau = 0$,

[35] Recall that the sequence (X_n) of random quantities is said to converge to X, in the mean-square, if $\mathbf{E}|X_n - X|^2 \to 0$, as $n \to \infty$.

that is,

$$\lim_{\tau\to 0}\gamma_X(\tau)=\gamma_X(0).$$

In particular, signals with autocorrelation functions $\gamma_X(\tau)=e^{|\tau|}$ or $\gamma_X(\tau)=\frac{1}{1+\tau^2}$ are mean-square continuous.

A similar, mean-square analysis of the limit at $\tau=0$ of the differential ratio,

$$\mathbf{E}\left(\frac{X(t+\tau)-X(t)}{\tau}\right)^2=2\frac{\gamma_X(0)-\gamma_X(\tau)}{\tau^2},$$

shows that a stationary signal with autocorrelation function $\gamma_X(\tau)=e^{|\tau|}$ cannot be possibly mean-square differentiable because in this case

$$\lim_{\tau\to 0}\frac{\gamma_X(0)-\gamma_X(\tau)}{\tau^2}=\lim_{\tau\to 0}\frac{1-e^{-|\tau|}}{\tau^2}=\infty,$$

whereas the differentiability cannot be excluded for the signal with autocorrelation $\gamma_X(\tau)=\frac{1}{1+\tau^2}$ because in this case

$$\lim_{\tau\to 0}\frac{\gamma_X(0)-\gamma_X(\tau)}{\tau^2}=\lim_{\tau\to 0}\frac{1-\frac{1}{1+\tau^2}}{\tau^2}=1.$$

Of course, the above brief discussion just verifies the boundedness of the variance of the signal's differential ratio as $\tau\to 0$, not whether the latter has a limit. So, let us take a closer look at the issue of the mean-square differentiability of a stationary signal, that is the existence of the random quantity $X'(t)$, for a fixed t. First, observe that this existence is equivalent to the statement that[36]

$$\lim_{\tau_1\to 0}\lim_{\tau_2\to 0}\mathbf{E}\left(\frac{X(t+\tau_1)-X(t)}{\tau_1}-\frac{X(t+\tau_2)-X(t)}{\tau_2}\right)^2=0$$

But the expression under the limit signs is equal to

$$\mathbf{E}\left(\frac{X(t+\tau_1)-X(t)}{\tau_1}\right)^2+\mathbf{E}\left(\frac{X(t+\tau_2)-X(t)}{\tau_2}\right)^2 \\ -2\mathbf{E}\left(\frac{X(t+\tau_1)-X(t)}{\tau_1}\cdot\frac{X(t+\tau_2)-X(t)}{\tau_2}\right).$$

So, the existence of the derivative $X'(t)$ in the mean-square is equivalent to the fact that the first two terms converge to $\gamma_{X'}(0)$ and the third to

---

[36] This argument relies on the so-called Cauchy criterion of convergence for random quantities with finite variance: *A sequence X_n converges in the mean-square as $n\to\infty$, that is, there exists a random quantity X such that* $\lim_{n\to\infty}\mathbf{E}(X_n-X)^2=0$, *if and only if* $\lim_{n\to\infty}\lim_{m\to\infty}\mathbf{E}(X_n-X_m)^2=0$. This criterion permits the verification of the convergence without knowing what the limit is; see, e.g., Theorem 11.4.2 in W. Rudin, *Principles of Mathematical Analysis*, McGraw-Hill, New York, 1976.

$-2\gamma_{X'}(0)$. But the convergence of the last term means the existence of the limit

$$\begin{aligned}
&\lim_{\tau_1\to 0}\lim_{\tau_2\to 0}\frac{1}{\tau_1\tau_2}\mathbf{E}((X(t+\tau_1)-X(t))\cdot(X(t+\tau_2)-X(t)))\\
&=\lim_{\tau_1\to 0}\lim_{\tau_2\to 0}\frac{1}{\tau_1\tau_2}(\gamma_X(\tau_2-\tau_1)-\gamma_X(\tau_1)-\gamma_X(\tau_2)+\gamma_X(0))\\
&=\lim_{\tau_1\to 0}\lim_{\tau_2\to 0}\frac{1}{\tau_1\tau_2}\Delta_{-\tau_1}\Delta_{\tau_2}\gamma_X(0),
\end{aligned}$$

where $\Delta_\tau f(t) := f(t+\tau)-f(t)$ is the usual difference operator. Indeed,

$$\begin{aligned}
\Delta_{-\tau_1}\Delta_{\tau_2}\gamma_X(0) &= \Delta_{\tau_1}(\gamma_X(\tau_2)-\gamma_X(0))\\
&= (\gamma_X(\tau_2-\tau_1)-\gamma_X(-\tau_1))-(\gamma_X(\tau_2)-\gamma_X(0)).
\end{aligned}$$

Since the existence of the last limit appearing above means twice differentiability of the autocorrelation function of X at $\tau = 0$ we arrive at the following criterion:

A stationary signal $X(t)$ is mean-square differentiable if and only if its autocorrelation function $\gamma_X(\tau)$ is twice differentiable at $\tau = 0$. Moreover, in this case, the cross-correlation of the signal $X(t)$ and its derivative $X'(t)$ is

$$\mathbf{E}X(t)X'(s) = \lim_{\tau\to 0}\frac{\gamma_X(t+\tau-s)-\gamma_X(t-s)}{\tau} = \frac{\partial}{\partial t}\gamma_X(t-s), \tag{8.4.1}$$

and the autocorrelation of the derivative signal is

$$\mathbf{E}X'(t)X'(s) = \lim_{\tau\to 0}\left(\frac{\partial}{\partial t}\gamma_X(t+\tau-s)-\frac{\partial}{\partial t}\gamma_X(t-s)\right) = \frac{\partial^2}{\partial t\partial s}\gamma_X(t-s). \tag{8.4.2}$$

In a similar fashion one can calculate the cross-correlation of higher derivatives of the signal $X(t)$ to obtain that[37]

$$\mathbf{E}X^{(n)}(t)X^{(m)}(s) = \frac{\partial^{n+m}}{\partial t^n\partial s^m}\gamma_X(t-s). \tag{8.4.3}$$

Sample path continuity. A study of properties of the individual sample paths (trajectories, realizations) of stationary random signals is a more delicate matter, with the most precise results obtainable only in the case of Gaussian signals. Indeed, we have observed in the previous sections that, for a Gaussian signal, the autocorrelation function determines all the finite-dimensional probability distributions of the signal, meaning

[37] For details, see M. Loeve, *Probability Theory*, Van Nostrand, Princeton, NJ, 1963, Section 34.3.

that for any finite sequence of windows, $[a_1, b_1], [a_2, b_2], \ldots, [a_N, b_N]$, and any collection of time instants $t_1, t_2, \ldots, t_N$, we can find the probability that the signal fits into those windows at prescribed times, that is,

$$\mathbf{P}(a_1 < t_1 < b_1,\ a_2 < t_2 < b_2, \ldots,\ a_N < t_N < b_N).$$

So it seems that by taking N to ∞, and making the time instants closer to each other and the windows narrower, one could find the probability that the signal's sample path has any specific shape or property. This idea is, roughly speaking, correct, but only in a subtle sense that will be explained below.

The discussion of the sample path properties of stationary signals will be based here on the following theorem of the theory of general random signals (stochastic processes) due to N. N. Kolmogorov.

Theorem 8.4.1. *Let $g(h)$ be an even function, nondecreasing for $h > 0$, and such that $g(h) \to 0$ as $h \to 0$. Furthermore, suppose that $X(t)$ is a random signal such that*

$$\mathbf{P}(|X(t+h) - X(t)| > g(h)) \le q(h), \tag{8.4.4}$$

for a function $q(h)$ satisfying the following three conditions:

$$q(h) \to 0 \quad \text{as } h \to 0; \tag{8.4.5}$$

$$\sum_{n=1}^{\infty} 2^n q(2^{-n}) < \infty; \tag{8.4.6}$$

$$\sum_{n=1}^{\infty} g(2^{-n}) < \infty; \tag{8.4.7}$$

Then, with probability 1, *the sample paths of the signal $X(t)$ are continuous.*

Although the proof of the above theorem is beyond the scope of this book,[38] the intuitive meaning of the assumptions (8.4.4)–(8.4.7) is clear: for the signal to have continuous sample paths, the increments of the signal over small time intervals can be permitted to be large only with a very small probability.

Applied to the second-order (not necessarily stationary) signals, Theorem 8.4.1 immediately gives the following.

[38] For a more complete discussion of this theorem and its consequences for sample path continuity and differentiability of random signals, see, for example, M. Loève, *Probability Theory*, Van Nostrand, Princeton, NJ, 1963, Section 35.3.

Corollary 8.4.1. *If there exists a τ_0 such that, for all τ, $0 \le \tau < \tau_0$, and all t in a finite time interval,*

$$\mathbf{E}(X(t+\tau) - X(t))^2 \le C|\tau|^{1+\epsilon} \tag{8.4.8}$$

for some constants $C, \epsilon > 0$, then the sample paths of the signal $X(t)$ are continuous with probability 1.

To see how Corollary 8.4.1 follows from Theorem 8.4.1,[39] observe first that for any random quantity Z and any constant $a > 0$.

$$\mathbf{P}(Z > a) \le \int_a^\infty f_Z(z)\,dz \le \int_a^\infty \frac{z^2}{a^2} f_Z(z)\,dz \le \frac{\mathbf{E}Z^2}{a^2}.$$

Condition (8.4.8) then implies that

$$\mathbf{P}(X(t+\tau) - X(t)| > g(\tau)) \le \frac{C|\tau|^{1+\epsilon}}{g^2(\tau)},$$

so that by selecting $g(\tau) = |\tau|^{\epsilon/4}$, and

$$q(\tau) = \frac{C|\tau|^{1+\epsilon}}{g^2(\tau)} = C|\tau|^{1+\epsilon/2},$$

we easily see that $g(\tau)$ and $q(\tau)$ are continuous functions vanishing at $\tau = 0$, and that the conditions (8.4.6)–(8.4.7) of the theorem are also satisfied. Indeed,

$$\sum_{n=1}^\infty 2^n q(2^{-n}) = C\sum_{n=1}^\infty 2^n (2^{-n})^{1+\epsilon/2} = C\sum_{n=1}^\infty 2^{-n\epsilon/2} < \infty,$$

and

$$\sum_{n=1}^\infty g(2^{-n}) = \sum_{n=1}^\infty 2^{-n\epsilon/4} < \infty.$$

In the special case of a stationary signal we have $\mathbf{E}(X(t+\tau) - X(t))^2 = 2(\gamma_X(0) - \gamma_X(\tau))$, so the sample path continuity is guaranteed by the following condition on the autocorrelation function:

$$|\gamma_X(0) - \gamma_X(\tau)| \le C|\tau|^{1+\epsilon}, \tag{8.4.9}$$

[39] This inequality is known as the Chebyshev inequality, and its proof here has been carried out only in the case of absolutely continuous probability distributions. The proof in the discrete case is left to the reader as an exercise; see Section 8.5.

for some constant $\epsilon > O$, and small enough τ.

In particular, for the autocorrelation function $\gamma_X(\tau) = \frac{1}{1+\tau^2}$,

$$|\gamma_X(0) - \gamma_X(\tau)| = 1 - \frac{1}{1+\tau^2} = \frac{\tau^2}{1+\tau^2} \le \tau^2,$$

and the condition (8.4.8) is satisfied, thus giving the sample path continuity.

However, for a signal with autocorrelation function $\gamma_X(\tau) = e^{-|\tau|}$, the difference $\gamma_X(0) - \gamma_X(\tau)$ behaves asymptotically like τ for $\tau \to 0$. Therefore, there is no positive ϵ for which condition (8.4.9) is satisfied and we cannot claim the continuity of the sample path in this case—not a surprising result if one remembers that the exponential autocorrelation was first encountered in the context of the obviously sample path discontinuous switching signal. Nevertheless, as we observed at the beginning of this section, a signal with an exponential autocorrelation is mean-square continuous.

For a Gaussian stationary signal $X(t)$, Theorem 8.4.1 can be applied in a more precise fashion since the probabilities $\mathbf{P}(X(t+\tau) - X(t) > a)$ are known exactly. Indeed, since for any positive z,

$$\int_z^\infty e^{-x^2/2}dx \le \int_z^\infty \frac{x}{z}e^{-x^2/2}dx = \frac{1}{z}e^{-z^2/2},$$

because $\frac{x}{z} \ge 1$ in the interval of integration, we have, for any nonnegative function $g(\tau)$ and positive constant C,

$$\mathbf{P}(|X(t+\tau) - X(t)| > Cg(\tau)) \le \sqrt{\frac{2}{\pi}}\frac{\sigma(\tau)}{Cg(\tau)} \exp\left(-\frac{1}{2}\frac{C^2g^2(\tau)}{\sigma^2(\tau)}\right), \tag{8.4.10}$$

where $\sigma^2(\tau) = \mathbf{E}(X(t+\tau) - X(t))^2 = 2(\gamma_X(0) - \gamma_X(\tau))$. This estimate yields the following result.

Corollary 8.4.2. *If there exists τ_0 such that, for all τ, $0 \le \tau \le \tau_0$, the autocorrelation function $\gamma_X(\tau)$ of a stationary Gaussian signal $X(t)$ satisfies the condition*

$$\gamma_X(0) - \gamma_X(\tau) \le \frac{K}{|\ln|\tau||^\delta}, \tag{8.4.11}$$

for some constants $K > 0$ and $\delta > 3$, then the signal $X(t)$ has continuous sample paths with probability 1.

The proof of the corollary is completed by selecting

$$g(\tau) = |\ln|\tau||^{-\nu},$$

with any number ν satisfying condition $1 < \nu < \frac{\delta-1}{2}$, choosing

$$q(C,\tau) = \frac{K'}{C|\ln|\tau||^{\delta/2-\nu}} \exp\left(-\frac{C^2}{2K}|\ln|\tau||^{\delta-2\nu}\right)$$

and verifying the convergence of the two series in conditions (8.4.6)–(8.4.7); see the exercise in Section 8.5.

Returning to the case of a stationary random signal with an exponential autocorrelation function, we see that if the signal is Gaussian, then Corollary 8.4.2 guarantees the continuity of its sample paths with probability 1. Indeed, condition (8.4.11) is obviously satisfied since (e.g., picking $\delta = 4$) we have

$$\lim_{\tau\to 0}(\gamma_X(0) - \gamma_X(\tau)) \cdot |\ln|\tau||^4 = \lim_{\tau\to 0}(1 - e^{-|\tau|}) \cdot |\ln|\tau||^4 = 0$$

in view of l'Hospital's rule.

8.5 Problems and exercises

8.5.1. A zero-mean Gaussian random signal has the autocorrelation function of the form

$$\gamma_X(\tau) = 10e^{-0.1|\tau|}\cos 2\pi\tau.$$

Write the covariance matrix for the signal sampled at four time instants separated by 0.5 seconds.

8.5.2. Find the joint p.d.f. of the signal from Exercise 8.5.1 at $t_1 = 1$ and $t_2 = 2.5$. Write the integral formula for

$$P(0 \le X(1) \le 1, 0 \le X(2.5) \le 2).$$

Evaluate the above probability numerically.

8.5.3. Find the joint p.d.f. of the signal from Exercise 8.5.1 at $t_1 = 1$, $t_2 = 1.5$, and $t_3 = 2.5$. Write the integral formula for

$$P(0 \le X(1) \le 1, -1 \le X(1.5) \le 3, 0 \le X(2.5) \le 2).$$

Evaluate the above probability numerically.

8.5.4. Show that if a 2D Gaussian random vector $\vec{Y} = (Y_1, Y_2)$ has uncorrelated components Y_1, Y_2, then those components are statistically independent random quantities.

8.5.5. Produce 2D surface plots for p.d.f.s of three Gaussian random vectors: $(X(1.0), X(1.1))^T$, $(X(1.0), X(2.0))^T$, $(X(1.0), X(5.0))^T$, where $X(t)$ is the stationary signal described in Example 8.3.1. Comment on similarities and differences in the three plots.

8.5.6. Prove that if there exists a τ_0 such that, for all $\tau < \tau_0$ and all t in a finite time interval,

$$\mathbf{E}(X(t+\tau) - X(t))^2 \le C \frac{|\tau|}{|\ln|\tau||^{1+\delta}},$$

for some $C > 0$ and $\delta > 2$, then the sample paths of the signal $X(t)$ are continuous with probability 1. *Hint*: This result is a little more delicate than Corollary 8.4.1, but the idea of the proof is similar: take $g(\tau) = |\ln|\tau||^{-\beta}$, for a β between 1 and $\frac{\delta}{2}$, from which we have

$$q(\tau) = \frac{|\tau|}{|\ln|\tau||^{1+\delta-2\beta}},$$

and check conditions (8.4.4)-(8.4.7) in Theorem 8.4.1.

8.5.7. Verify the Chebyshev inequality, $\mathbf{P}(|Z| > a) \le \frac{\mathbf{E}Z^2}{a^2}$, $a > 0$, for a discrete random quantity Z.

8.5.8. Produce plots of several 2D Gaussian densities with selected means and covariance matrices.

8.5.9. Random signal $X(t)$ has an autocorrelation function of the form $\gamma_X(\tau) = \exp(-|\tau|^\alpha)$ with $0 < \alpha \le 2$. For which values of parameter α can you claim the continuity of sample paths of $X(t)$ with probability 1?

8.5.10. Verify formula (8.4.3) for the cross-correlation of higher derivatives of a stationary signal.

8.5.11. Verify the convergence of the series (8.4.6)-(8.4.7) in the proof of Corollary 8.4.2.

9

Discrete Signals and Their Computer Simulations

Given an arbitrary power spectrum, our ability to simulate the corresponding stationary random signals, using only the random number generator which produces, say, discrete white noise, depends on the observation that in some sense all stationary random signals can be approximated by superpositions of random harmonic oscillations descirbed in Example 4.1.2. The observation itself is not obvious at all and, of course, the key to applying it is in the details: In what is sense the approximation meant? What is the precise algorithm for obtaining such an approximation?

In this chapter we work with discrete-time signals, and the rigorous answer to the above questions is contained in the so-called *spectral representation theorem* for stationary random signals which is derived in this chapter. On the way to its formulation we introduce the necessary concepts including the crucial construction of *stochastic integrals* with respect to a white noise signal, often called the *white noise integrals*. We conclude with a computer algorithm based on the spectral representation theorem.

9.1 Autocorrelation as a positive definite function

In this chapter we will study random stationary signals in discrete time, that is, sequences of *complex-valued* random quantities

$$\dots, X(-2), X(-1), X(0), X(1), X(2), \dots$$

with time n extending all the way from minus to plus infinity. The stationarity is meant in the second-order, weak sense, that is, we will assume that the means $\mathbf{E}X(n) = 0$ and the autocorrelation function, now really a sequence,

$$\mathbf{E}X(n)X^*(m) = \gamma(n-m), \quad m, n = \dots, -2, -1, 0, 1, 2, \dots,$$

depends only on the time lag $n - m$. The following properties of the autocorrelation sequence are immediately verified:

For any n,

$$\mathbf{E}|X(n)|^2 = \mathbf{E}X(n)[X(n)]^* = \mathbf{E}|X(0)|^2 = \gamma_X(0) \geq 0, \tag{9.1.1}$$

$$\gamma_X(-n) = \gamma_X^*(n), \tag{9.1.2}$$

$$|\gamma_X(n)| \leq \gamma_X(0). \tag{9.1.3}$$

The last inequality is a direct consequence of the Cauchy-Schwartz inequality.

Also, importantly, the autocorrelation sequence is *positive definite*, that is, for any positive integer N, arbitrary integers, $n_1, n_2, \ldots, n_N$, and arbitrary complex numbers $\lambda_1, \lambda_2, \ldots, \lambda_N$,

$$\sum_{i,k=1}^{N} \gamma_X(n_i - n_k)\lambda_i\lambda_k^* \geq 0. \tag{9.1.4}$$

Indeed,

$$\begin{aligned}
\sum_{i,k=1}^{N} \gamma_X(n_i - n_k)\lambda_i\lambda_k^* &= \sum_{i,k=1}^{N} \mathbf{E}[X(n_i)X^*(n_k)]\lambda_i\lambda_k^* \\
&= \mathbf{E} \sum_{i,k=1}^{N} [\lambda_i X(n_i)] \cdot [\lambda_k X(n_k)]^* \\
&= \mathbf{E} \sum_{i=1}^{N} \lambda_i X(n_i) \cdot \sum_{k=1}^{N} [\lambda_k X(n_k)]^* \\
&= \mathbf{E} \left| \sum_{i=1}^{N} \lambda_i X(n_i) \right|^2 \geq 0.
\end{aligned}$$

9.2 Cumulative power spectrum of discrete-time stationary signal

The development of this section will be analogous to the development of the concept of the power spectrum of continuous-time signals in Section 5.2. However, we will proceed in a slightly different fashion, and with more mathematical precision. The basic structural result regarding the autocorrelation function of a discrete-time stationary signal can be formulated as follows:

Herglotz Theorem. *The following statements about sequence* $\gamma(n)$, $n = \ldots, -2, -1, 0, 1, 2, \ldots$, *of complex numbers are equivalent:*

(i) *Sequence $\gamma(n)$ is an autocorrelation sequence of a stationary discrete-time signal, that is, there exists a stationary signal $X(n)$ such that $\gamma(n) = \gamma_X(n)$.*

(ii) *The sequence $\gamma(n)$ is positive definite, that is, it satisfies condition (9.1.4).*

(iii) *There exists a nondecreasing bounded function $\mathfrak{S}_X(f)$, defined on the interval $[0,1]$, such that*

$$\gamma(n) = \int_0^1 e^{j2\pi nf}\, d\mathfrak{S}(f), \quad n = \ldots, -2, -1, 0, 1, 2, \ldots. \tag{9.2.1}$$

The function $\mathfrak{S}_X(f)$ is called the cumulative power spectrum of signal X.

Remark 9.2.1 (*power spectrum density*). The integral of the form

$$\int a(f) d\mathfrak{S}(f),$$

called the Stieltjes integral, is to be understood as the limit of sums $\sum a(f_i) \cdot \Delta\mathfrak{S}(f_i)$, with $\Delta\mathfrak{S}(f_i) = \mathfrak{S}(f_i) - \mathfrak{S}(f_{i-1})$.

If the cumulative power spectrum has a spectral density $S(f)$, $0 \le f \le 1$, that is,

$$\mathfrak{S}(f) = \int_0^f S(g) dg, \qquad \frac{d\mathfrak{S}(f)}{df} = S(f) \ge 0,$$

then formula (9.1.1) takes the form of the usual Riemann integral

$$\gamma(n) = \int_0^1 e^{j2\pi nf} S(f) df, \quad n = \ldots, -2, -1, 0, 1, 2, \ldots, \tag{9.2.2}$$

and the sequence $\gamma(-n)$ can be simply viewed as the sequence of Fourier coefficients of power spectrum density $S(f)$.

In the special case when the cumulative power spectrum is constant, except for jumps, that is,

$$\mathfrak{S}(f) = \sum_l s_l u(f - f_l),$$

where $u(t)$ is the unit step function, then

$$\int a(f) dS(f) = \sum_l a(f_l) s_l,$$

so that

$$\gamma(n) = \sum_l s_l e^{j2\pi n f_l}, \quad n = \dots, -2, -1, 0, 1, 2, \dots \tag{9.2.3}$$

and the power spectrum density can be understood as a sum of the Dirac deltas.

$$S(f) = \sum_l s_l \delta(f - f_l).$$

However, it is worth remembering that there are so-called singular cumulative power spectra that are not of either of the two types described above (nor their mixtures).[40]

The implication (i) $\Longrightarrow$ (ii) has been proved following the definition (9.1.4).

We shall now prove that (ii) $\Longrightarrow$ (iii). So, assume that $\gamma(n)$ is positive definite. In view of (9.1.4), selecting $n_i = i, \lambda_i = e^{-j2\pi i f}, i = 1, 2, \dots, N$, we have

$$\begin{aligned} 0 \le \sum_{i,k=1}^{N} \gamma(i-k) e^{-j2\pi i f} e^{j2\pi k f} &= \sum_{i,k=1}^{N} \gamma(i-k) e^{-j2\pi(i-k)f} \\ &= \sum_{m=-N+1}^{N-1} (N - |m|)\gamma(m) e^{-j2\pi m f}, \end{aligned}$$

after substitution $m = i - k$. Define

$$S_N(f) := \frac{1}{N} \sum_{m=-N+1}^{N-1} (N - |m|)\gamma(m) e^{-j2\pi m f}.$$

Then

$$S_N(f) \ge 0 \quad \text{and} \quad \int_0^1 S_N(f)\,df = \gamma(0). \tag{9.2.4}$$

By a fundamental real analysis result called the Arzela-Ascoli theorem,[41] conditions (9.2.4) guarantee the existence of a function $\mathfrak{S}(f)$ and a sequence $N_i \nearrow \infty, i \to \infty$ such that, for each bounded and smooth function $a(f)$,

$$\int_0^1 a(f) S_{N_i}(f)\,df \longrightarrow \int_0^1 a(f)\,d\mathfrak{S}(f).$$

Therefore, selecting $a(f) = e^{j2\pi m f}$, we have

$$\int_0^1 e^{j2\pi m f}\,d\mathfrak{S}(f) = \lim_{i\to\infty} \int_0^1 e^{j2\pi m f} S_{N_i}(f)\,df = \gamma(m)$$

40 See Section 3.1 or, e.g., M. Denker and W. A. Woyczyński, *Introductory Statistics and Random Phenomena: Uncertainty, Compexity, and Chaotic Behavior in Engineering and Science*, Birkhäuser Boston, Cambridge, MA, 1998.

41 See, e.g., G. B. Folland, *Real Analysis*, Wiley, New York, 1984.

because, for each m such that $|m| \le N_i$,

$$\int_0^1 e^{j2\pi m f} S_{N_i}(f)\,df = \gamma(m)\left(1 - \frac{|m|}{N_i}\right).$$

Thus the existence of the cumulative spectral measure for each discrete-time stationary signal has been established.

The implication (iii) $\Longrightarrow$ (ii) can be verified directly. Indeed,

$$\begin{aligned}\sum_{i,k=1}^{N} \gamma(n_i - n_k)\lambda_i\lambda_k^* &= \sum_{i,k=1}^{N} \int_0^1 e^{j2\pi(n_i-n_k)f}\,d\mathfrak{S}(f)\cdot\lambda_i\lambda_k^* \\ &= \int_0^1 \sum_{i,k=1}^{N} [\lambda_i e^{j2\pi n_i f}]\cdot[\lambda_k e^{j2\pi n_k f}]^*\,d\mathfrak{S}(f) \\ &= \int_0^1 \left|\sum_{i=1}^{N} \lambda_i e^{j2\pi n_i f}\right|^2 d\mathfrak{S}(f) \ge 0.\end{aligned}$$

The implication (ii) $\Longrightarrow$ (i) follows from the following fact established in Section 8.2. For any given positive definite matrix $\boldsymbol{\Gamma} = (\gamma_{ik})$, $i,k = 1,2,\ldots,N$, there exists a Gaussian random vector $\mathbf{X} = (X_1, X_2, \ldots, X_N)$, with covariance matrix $\boldsymbol{\Gamma}$. Now, for any N, it suffices to take $\boldsymbol{\Gamma} = (\gamma(i-k))$, $i,k = 1,2,\ldots,N$, and define

$$X(1) = X_1, \quad X(2) = X_2, \quad \ldots, \quad X(N) = X_N.$$

This proves the existence of a finite discrete-time stationary random signal with an autocorrelation sequence given by a prescribed positive definite sequence.[42]

9.3 Stochastic integration with respect to signals with uncorrelated increments

Recall that our goal in this chapter is to develop a simulation algorithm for discrete-time stationary signals with a given power spectrum, and one of the methods used for that purpose involves representation of the random signal as a stochastic integral with respect to another random signal which has independent increments which is easy to simulate. The purpose of this section is to introduce the latter concept.

The finite variance, zero-mean real-valued signal $\mathcal{W}(w)$ of continuous or discrete parameter w is said to have uncorrelated increments if, for any $w_1 \le w_2 \le w_3$,

[42] A step proving the existence of an *infinite* such sequence requires an application of the so-called Kolmogorov extension theorem; see, e.g., P. Billingsley, *Probability and Measure*, Wiley, New York, 1986.

$$\mathbf{E}[(\mathcal{W}(w_3) - \mathcal{W}(w_2)) \cdot (\mathcal{W}(w_2) - \mathcal{W}(w_1))] = 0. \tag{9.3.1}$$

In other words, such signals have uncorrelated increments over disjoint intervals of parameter w. Observe that condition (9.3.1) can be rewritten in terms of the autocorrelation function $\gamma_{\mathcal{W}}(v,w) = \mathbf{E}\mathcal{W}(v)\mathcal{W}(w)$ (which here is truly a function of two variables v, w, and not just the parameter lag $w - v$ as is the case for stationary signals) as follows:

$$\begin{aligned}
&\mathbf{E}[(\mathcal{W}(w_3) - \mathcal{W}(w_2)) \cdot (\mathcal{W}(w_2) - \mathcal{W}(w_1))] \qquad (9.3.2)\\
&= \mathbf{E}\mathcal{W}(w_3)\mathcal{W}(w_2) - \mathbf{E}\mathcal{W}(w_2)\mathcal{W}(w_2) - \mathbf{E}\mathcal{W}(w_3)\mathcal{W}(w_1)\\
&\quad + \mathbf{E}\mathcal{W}(w_2)\mathcal{W}(w_1)\\
&= \gamma_{\mathcal{W}}(w_3,w_2) - \gamma_{\mathcal{W}}(w_2,w_2) - \gamma_{\mathcal{W}}(w_3,w_1) + \gamma_{\mathcal{W}}(w_2,w_1) = 0.
\end{aligned}$$

Example 9.3.1 (the cumulative white noise in discrete time). In discrete time, the white noise $W(n)$ was defined simply as a sequence of zero-mean uncorrelated random quantities with equal variance, that is,

$$\gamma_W(n,m) = \gamma_W(m-n) = \mathbf{E}W(n)W(m) = \begin{cases} 0 & \text{if } n - m \neq 0; \\ \sigma^2 & \text{if } n - m = 0. \end{cases}$$

We will define the *cumulative white noise* as the random signal

$$\mathcal{W}(n) = W(1) + W(2) + \cdots + W(n), \quad n = 1, 2, \ldots,$$

with the convention $\mathcal{W}(0) = 0$. The cumulative white noise has uncorrelated increments. Indeed, if $n_1 \le n_2 \le n_3$, then

$$\begin{aligned}
&\mathbf{E}[(\mathcal{W}(n_3) - \mathcal{W}(n_2)) \cdot (\mathcal{W}(n_2) - \mathcal{W}(n_1))]\\
&= \mathbf{E}\left[\left(\sum_{n=1}^{n_3} W(n) - \sum_{n=1}^{n_2} W(n)\right) \cdot \left(\sum_{n=1}^{n_2} W(n) - \sum_{n=1}^{n_1} W(n)\right)\right]\\
&= \mathbf{E}[(W(n_2+1) + \cdots + W(n_3)) \cdot (W(n_1+1) + \cdots + W(n_2))]\\
&= 0.
\end{aligned}$$

For any signal $\mathcal{W}(w)$ with uncorrelated increments, we will introduce a *cumulative control function*

$$C(w) := \mathbf{E}[\mathcal{W}(w) - \mathcal{W}(0)]^2 \ge 0. \tag{9.3.3}$$

which simply measures the variance of the increment of the signal from 0 to w. Since the variance of the sum of uncorrelated random quantities is the sum of their variances, the cumulative control function is always nondecreasing because, for $0 \le v \le w$,

$$\begin{aligned}
C(w) &= \mathbf{E}[(\mathcal{W}(w) - \mathcal{W}(0)]^2\\
&= \mathbf{E}[(\mathcal{W}(w) - \mathcal{W}(v)) + (\mathcal{W}(v) - \mathcal{W}(0))]^2
\end{aligned}$$

$$= \mathbf{E}[(\mathcal{W}(w) - \mathcal{W}(v))]^2 + \mathbf{E}[(\mathcal{W}(v) - \mathcal{W}(0))]^2$$
$$\geq \mathbf{E}[(\mathcal{W}(v) - \mathcal{W}(0))]^2 = C(v). \tag{9.3.4}$$

Observe that, under condition $\mathcal{W}(0) = 0$, the cumulative control function determines the correlation structure of $\mathcal{W}(w)$ and vice versa. If, say, $0 \leq v \leq w$, then

$$\begin{aligned} \gamma_{\mathcal{W}}(v, w) &= \mathbf{E}\mathcal{W}(v)\mathcal{W}(w) \\ &= \mathbf{E}[\mathcal{W}(v) - \mathcal{W}(0)] \cdot [(\mathcal{W}(w) - \mathcal{W}(v)) + (\mathcal{W}(v) - \mathcal{W}(0)] \\ &= \mathbf{E}[\mathcal{W}(v) - \mathcal{W}(0)] \cdot [(\mathcal{W}(v) - \mathcal{W}(0)] = C(v) \end{aligned}$$

because the increments over intervals $[0, v]$ and $[v, w]$ are uncorrelated. Since an analogous reasoning holds true in the case $0 \leq w \leq v$, we get the general formula

$$\gamma_{\mathcal{W}}(v, w) = C(\min(v, w)). \tag{9.3.5}$$

An important class of signals with uncorrelated increments are those that also have *stationary increments*, that is for which the c.d.f. of the increment $\mathcal{W}(w) - \mathcal{W}(v)$ is the same as the c.d.f. of the increment $\mathcal{W}(w + z) - \mathcal{W}(v + z)$, for any z. In this case, the cumulative control function satisfies the condition

$$C(w + v) = C(w) + C(v) \tag{9.3.6}$$

because $\mathbf{E}[\mathcal{W}(w + v) - \mathcal{W}(0)]^2 = \mathbf{E}[\mathcal{W}(w + v) - \mathcal{W}(v)]^2 + \mathbf{E}[\mathcal{W}(v) - \mathcal{W}(0)]^2 = \mathbf{E}[\mathcal{W}(w) - \mathcal{W}(0)]^2 + \mathbf{E}[\mathcal{W}(v) - \mathcal{W}(0)]^2$. Condition (9.3.6) forces the cumulative function to be of the form

$$C_{\mathcal{W}}(w) = \text{const} \cdot w, \tag{9.3.7}$$

and, in view of (9.2.4), the autocorrelation structure of a signal with stationary and uncorrelated increments is of the form

$$\gamma_{\mathcal{W}}(v, w) = \text{const} \cdot \min(v, w). \tag{9.3.8}$$

Example 9.3.2 (the Wiener process). A continuous-time Gaussian signal with stationary and independent increments with

$$C_{\mathcal{W}}(w) = w, \qquad \gamma_{\mathcal{W}}(v, w) = \min(v, w).$$

is called the Wiener stochastic process (or the Brownian motion process). Its sample trajectories are shown in Figure 1.1.4. Notice that in this case the autocorrelation function gives a complete description of all finite-dimensional distributions of $\mathcal{W}(w)$. Indeed, given parameter values

$$w_1 \leq w_2 \leq \cdots \leq w_N,$$

the random vector

$$(\mathcal{W}(w_1), \mathcal{W}(w_2), \ldots, \mathcal{W}(w_N))$$

is a Gaussian random vector with the covariance matrix

$$\Gamma = (\min(w_i, w_k)),$$

so that its joint c.d.f. can be explicitly calculated:

$$\begin{aligned} \mathbf{P}(\mathcal{W}(w_1) \le a_1, \mathcal{W}(w_2) \le a_2, \ldots, \mathcal{W}(w_N) \le a_N) \qquad\qquad (9.3.9) \\ = \int_{-\infty}^{a_1}\int_{-\infty}^{a_2}\cdots\int_{-\infty}^{a_N} \frac{e^{-\zeta_1^2/2w_1}}{\sqrt{2\pi w_1}} \cdot \frac{e^{-(\zeta_2-\zeta_1)^2/2(w_2-w_1)}}{\sqrt{2\pi(w_2-w_1)}} \times \cdots \\ \times \cdots \times \frac{e^{-(\zeta_N-\zeta_{N-1})^2/2(w_N-w_{N-1})}}{\sqrt{2\pi(w_N-w_{N-1})}} \times d\zeta_N \cdots d\zeta_2 \cdot d\zeta_1. \end{aligned}$$

Now, we shall introduce the stochastic integral

$$\int_0^1 x(w)\,d\mathcal{W}(w),$$

with respect to a signal $\mathcal{W}(w)$ with uncorrelated increments, for a deterministic, possibly complex-valued function $x(w)$. If $x(w)$ is a step function of the form

$$x(w) = \sum_{i=1}^{N} x_i \mathbf{1}_{(w_{i-1}, w_i]}(w), \qquad (9.3.10)$$

with $0 = w_0 < w_1 < \cdots < w_{N-1} < w_N = 1$, and $\mathbf{1}_A(w)$ denoting the indicator function of set A,[43] then, obviously,

$$\int_0^1 x(w)\,d\mathcal{W}(w) := \sum_{i=1}^{N} x_i \cdot (\mathcal{W}(w_i) - \mathcal{W}(w_{i-1})). \qquad (9.3.11)$$

Note that the variance of this stochastic integral

$$\begin{aligned} \mathbf{E}\left|\int x(w)\,d\mathcal{W}(w)\right|^2 &= \mathbf{E}\left|\sum_{i=1}^{N} x_i \cdot (\mathcal{W}(w_i) - \mathcal{W}(w_{i-1}))\right|^2 \\ &= \sum_{i=1}^{N} |x_i|^2 \mathbf{E}(\mathcal{W}(w_i) - \mathcal{W}(w_{i-1}))^2 \\ &= \sum_{i=1}^{N} |x_i|^2 (C(w_i) - C(w_{i-1})) \end{aligned}$$

[43] Recall that the indicator function $\mathbf{1}_A(w)$ is defined as being equal to 1 for w belonging to set A and 0 for w outside A.

$$= \int_0^1 |x(w)|^2 dC(w) \tag{9.3.12}$$

because, in view of (9.3.4), for any $0 < v < w$,

$$\mathbf{E}(\mathcal{W}(w) - \mathcal{W}(v))^2 = C(w) - C(v). \tag{9.3.13}$$

Since any function $x(w)$ such that

$$\int_0^1 |x(w)|^2 dC(w) < \infty \tag{9.3.14}$$

is a limit of a sequence $x_n(w)$ of step functions,[44] in the sense that

$$\int_0^1 |x_n(w) - x(w)|^2 dC(w) \to 0 \quad \text{as } n \to \infty,$$

the definition (9.3.11) of the stochastic integral for step functions can now be extended to any $x(w)$ satisfying condition (9.3.14) by setting

$$\int_0^1 x(w) d\mathcal{W}(w) := \lim_{n\to\infty} \int_0^1 x_n(w) d\mathcal{W}(w), \tag{9.3.15}$$

where the limit is understood as the limit in the mean-square of random quantities. In view of this procedure, the general stochastic integral for a function $x(w)$ satisfying condition (9.3.14) enjoys the "isometric" property

---

$$\mathbf{E}\left|\int_0^1 x(w) d\mathcal{W}(w)\right|^2 = \int_0^1 |x(w)|^2 dC(w). \tag{9.3.16}$$

Example 9.3.3 (Gaussian stochastic integrals). Note that if the cumulative control function $C(w)$ of a Gaussian process with independent increments $\mathcal{V}(w)$ has a density $c(w)$, that is,

$$C(w) = \int_0^w c(v)dv, \quad \frac{dC(w)}{dw} = c(w) \geq 0, \quad 0 \leq w \leq 1,$$

then, in view of (9.3.16),

$$\mathbf{E}(\mathcal{V}(w))^2 = \mathbf{E}\left(\int_0^w d\mathcal{V}(v)\right)^2 = \int_0^w c(v)dv = \int_0^w \left(\sqrt{c(v)}\right)^2 dv,$$

which implies that, for any $x(w)$ satisfying (9.3.14), the statistical properties of the stochastic integrals

---

[44] See, e.g., G. B. Folland, *Real Analysis*, Wiley, New York, 1984.

$$\int_0^1 x(v)\,d\mathcal{V}(v) \quad \text{and} \quad \int_0^1 x(w)\sqrt{c(w)}\,d\mathcal{W}(w), \tag{9.3.17}$$

where $\mathcal{W}(w)$ is the Wiener process, are the same. Later on this fact will serve as the basis for computer simulation of stationary random signals with a given spectrum.

Because, for any complex numbers ξ, η, we have the so-called "polarization formulas,"

$$\mathrm{Re}[\xi \cdot \eta^*] = \frac{1}{4}(|\xi + \eta|^2 - |\xi - \eta|^2),$$

$$\mathrm{Im}[\xi \cdot \eta^*] = \frac{1}{4}(|\xi + j\eta|^2 - |\xi - j\eta|^2),$$

which express the product in terms of the squared moduli, the "isometric" relation (9.3.16) extends from the mean-squares to scalar products. In other words, for any $x(w)$, $y(w)$, satisfying condition (9.3.14), we have

$$\mathbf{E}\left[\int_0^1 x(w)\,d\mathcal{W}(w) \cdot \left(\int_0^1 y(w)\,d\mathcal{W}(w)\right)^*\right] = \int_0^1 x(w) \cdot y^*(w)\,dC(w). \tag{9.3.18}$$

9.4 Spectral representation of stationary signals

The fundamental result about the structure of discrete-time stationary signals is that they are, essentially, sequences of random Fourier coefficients of stochastic processes with uncorrelated increments. More precisely, we have the following.

Spectral Representation Theorem. *A discrete-time random signal $X(n)$, $n = \ldots, -2, -1, 0, 1, 2, \ldots$, is stationary if and only if it has the representation*

$$X(n) = \int_0^1 e^{j2\pi nf}\,d\mathcal{W}(f) \tag{9.4.1}$$

for a certain random process $\mathcal{W}(f)$, $0 \le f \le 1$, which has uncorrelated increments. Moreover, the cumulative spectral function of $X(n)$ is identical to the cumulative control function of $\mathcal{W}(f)$, that is,

$$\mathfrak{S}_X(f) = C_{\mathcal{W}}(f), \quad 0 \le f \le 1. \tag{9.4.2}$$

Proof. If random signal $X(n)$ is of the form (9.4.1), then it is stationary because it has zero mean and because, in view of the "isometry" (9.3.17),

$$\begin{aligned} \mathbf{E}[X(n)X^*(m)] &= \mathbf{E}\left[\int_0^1 e^{j2\pi nf}\,d\mathcal{W}(f)\cdot\left(\int_0^1 e^{j2\pi mf}\,d\mathcal{W}(f)\right)^*\right] \\ &= \int_0^1 e^{j2\pi(n-m)f}\,dC_{\mathcal{W}}(f). \end{aligned}$$

The above calculation also identifies the cumulative control function of process $\mathcal{W}(f)$ as the cumulative spectral function of the random signal $X(n)$.

The proof of the reverse implication is more delicate as it requires identification, for each signal $X(n)$, of a process $\mathcal{W}(f)$ yielding representation (9.4.1). Therefore, assume that $X(n)$ is a stationary signal with autocorrelation sequence

$$\gamma_X(n) = \int_0^1 e^{j2\pi nf}\,d\mathfrak{S}_X(f).$$

Denote by $L_0^2(\mathbf{P})$ the space of random quantities with zero mean and finite variance and by $L^2(d\mathfrak{S}_X(f))$ the space of complex functions on $[0,1]$ which are square integrable with respect to the cumulative spectral function $\mathfrak{S}_X(f)$. Next, consider a linear mapping I from $L_0^2(\mathbf{P})$ into $L^2(d\mathfrak{S}_X(f))$ defined by the identity

$$I[X(n)] := e^{j2\pi nf}, \quad n = \dots,-2,-1,0,1,2,\dots, \tag{9.4.3}$$

on complex exponentials and extended, in a natural way, to all their combinations. In other words, for any complex numbers $c_{-N},\dots,$ $c_{-1},c_0,c_1,\dots,c_N$,

$$I\left[\sum_{n=-N}^{N} c_n X(n)\right] = \sum_{n=-N}^{N} c_n e^{j2\pi nf}. \tag{9.4.4}$$

Mapping I is an *isometry*[45] on such linear combinations because

$$\begin{aligned} \mathbf{E}\left|\sum_{n=-N}^{N} c_n X(n)\right|^2 &= \sum_{n,m=-N}^{N} c_n c_m^* \mathbf{E}[X(n)X^*(m)] \\ &= \sum_{n,m=-N}^{N} c_n c_m^* \int_0^1 e^{j2\pi(n-m)f}\,d\mathfrak{S}_W(f) \\ &= \int_0^1 \left|\sum_{n=-N}^{N} c_n e^{j2\pi nf}\right|^2 d\mathfrak{S}_X(f), \end{aligned}$$

[45] In the sense that it preserves the norms: the standard deviation in space $L_0^2(\mathbf{P})$ and $\|a\| = (\int_0^1 |a(f)|^2 d\mathfrak{S}_X(f))^{1/2}$ for an $a(f)$ in $L^2(d\mathfrak{S}_X(f))$.

and, as such, it extends to the linear isometry

$$I : \mathcal{L}[X(n),\ n = \dots,-2,-1,0,1,2,\dots] \longmapsto L^2(d\mathfrak{S}_X(f)),$$

where $\mathcal{L}[X(n), n = \dots,-2,-1,0,1,2,\dots]$ is the subspace of $L^2(\mathbf{P})$ consisting of linear combinations of $X(n)$s and their mean-square limits. Since any isometry is necessarily a one-to-one mapping, I has a well-defined inverse

$$I^{-1} : L^2(d\mathfrak{S}_X(f)) \longmapsto \mathcal{L}[X(n),\ n = \dots,-2,-1,0,1,2,\dots],$$

which is also a linear isometry.

Now we will define a stochastic process $\mathcal{W}(f)$ by the formula

$$\mathcal{W}(f) := I^{-1}(\mathbf{1}_{[0,f]}),$$

where $\mathbf{1}_{[0,f]}(g), 0 \le g \le 1$, is the indicator function of the interval $[0, f]$. This process has zero mean and uncorrelated increments since, for $f_1 \le f_2 \le f_3$, in view of the isometric property of I^{-1},

$$\begin{aligned}
&\mathbf{E}[(\mathcal{W}(f_3) - \mathcal{W}(f_2)) \cdot (\mathcal{W}(f_2) - \mathcal{W}(f_1))] \\
&= \mathbf{E}[(I^{-1}(\mathbf{1}_{[0,f_3]}) - I^{-1}(\mathbf{1}_{[0,f_2]})) \cdot (I^{-1}(\mathbf{1}_{[0,f_2]}) - I^{-1}(\mathbf{1}_{[0,f_1]}))] \\
&= \mathbf{E}[I^{-1}(\mathbf{1}_{[0,f_3]}) - \mathbf{1}_{[0,f_2]})) \cdot (I^{-1}(\mathbf{1}_{[0,f_2]}) - \mathbf{1}_{[0,f_1]}))] \\
&= \mathbf{E}[I^{-1}(\mathbf{1}_{(f_2,f_3]}) \cdot (I^{-1}(\mathbf{1}_{(f_1,f_2]})] \\
&= \int_0^1 \mathbf{1}_{(f_2,f_3]}(f) \cdot \mathbf{1}_{(f_1,f_2]}(f)\, d\mathfrak{S}_X(f) = 0.
\end{aligned}$$

The same calculation shows that

$$\mathbf{E}\mathcal{W}^2(f) = \int_0^1 \mathbf{1}^2_{[0,f]}(g)\, d\mathfrak{S}_X(g) = \mathfrak{S}_X(f),$$

so the control function $C_{\mathcal{W}}(f) = \mathfrak{S}_X(f)$. Now, proceeding again via step functions as in Section 9.2, using the linearity and isometry properties of I^{-1}, we have for any function $a(f)$ in space $L^2(dC_X(f))$,

$$I^{-1}(a) = \int_0^1 a(f)\, d\mathcal{W}(f).$$

In particular, selecting $a(f) = e^{j2\pi nf}$, we obtain that

$$X(n) = I^{-1}(e^{j2\pi nf}) = \int_0^1 e^{j2\pi nf}\, d\mathcal{W}(f),$$

which concludes the proof of the spectral representation theorem. □

Example 9.4.1 (white noise signal). Let $\mathcal{W}(f)$ be the Wiener process. Its cumulative control function

$$C_{\mathcal{W}}(f) = f = \int_0^f df$$

has a control density function $c_{\mathcal{W}}(f) \equiv 1$. The stationary, discrete-time signal

$$X(n) = \int_0^1 e^{j2\pi nf}\, d\mathcal{W}(f)$$

has the spectral density function $S_X(f) = c_{\mathcal{W}}(f) \equiv 1$ and the autocorrelation sequence

$$\gamma_X(n) = \mathbf{E}X(n)X^*(0) = \int_0^1 e^{j2\pi nf}\, df = \delta(n) = \begin{cases} 0 & \text{if } n \neq 0; \\ 1 & \text{if } n = 0. \end{cases}$$

Hence $X(n)$ is the discrete-time white noise discussed in Chapter 5.

Example 9.4.2 (filtered white noise). Let $X(n)$ be the white noise discussed above. Consider the (acausal) filtered white noise signal

$$Y(n) = \sum_{k=-\infty}^{\infty} c_k X(n-k) = \int_0^1 \left(\sum_{k=-\infty}^{\infty} c_k e^{j2\pi(n-k)f} \right) d\mathcal{W}(f)$$

for $n = \ldots, -2, -1, 0, 1, 2, \ldots$. Its autocorrelation sequence is

$$\gamma_Y(n) = \mathbf{E}Y(n)Y^*(0) = \mathbf{E}\left(\sum_{k=-\infty}^{\infty} c_k X(n-k) \cdot \sum_{k=-\infty}^{\infty} c_k^* X^*(-k) \right)$$

$$= \mathbf{E} \sum_{k,l=-\infty}^{\infty} c_k c_l^* X(n-k) X^*(-l) = \sum_{k,l=-\infty}^{\infty} c_k c_l^* \delta(n-(k-l))$$

$$= \sum_{k,l=-\infty}^{\infty} c_k c_l^* \int_0^1 e^{j2\pi(n-(k-l))f}\, df = \int_0^1 |c(f)|^2 e^{j2\pi nf}\, df,$$

where

$$c(f) = \sum_{k=-\infty}^{\infty} c_k e^{-j2\pi kf},$$

is well defined as long as $\sum_{k=-\infty}^{\infty} |c_k|^2 < \infty$. Hence the power spectral density of the filtered white noise is

$$S_Y(f) = |c(f)|^2.$$

9.5 Computer algorithms

Given a spectral density $S_X(f)$ of a discrete-time, stationary Gaussian signal $X(n)$, we can simulate a sample path of $X(n)$, $n = 1, 2, \ldots, N$, by first calculating the autocorrelation function $\gamma_X(n)$ using formula (9.2.2),

$$\gamma_X(n) = \int_0^1 e^{j2\pi nf} S_X(f)\,df, \tag{9.5.1}$$

and then by producing a sample of an N-dimensional Gaussian random vector $\mathbf{X} = (X_1, X_2, \ldots, X_n)$ with the covariance matrix $\mathbf{\Gamma} = (\gamma_X(n - m), n, m = 1, 2, \ldots, N)$, using standard statistical software. This, however, would be computationally expensive, and even infeasible if n is large.

Therefore, in this section we will describe a different, explicit algorithm for such a simulation based on the spectral representation of Section 9.4. The algorithm is mathematically justified by the discussions of the preceding sections, and it has the advantage of not being restricted to Gaussian signals.

The starting point is, of course, the spectral representation theorem and, in particular, formula (9.4.1) which writes the signal $X(n)$ as a random Fourier coefficient,

$$X(n) = \int_0^1 e^{j2\pi nf}\, d\mathcal{W}(f), \quad n = 1, 2, \ldots, N, \tag{9.5.2}$$

of a process $\mathcal{W}(f)$ with uncorrelated increments and cumulative control function $C_{\mathcal{W}}(f)$ equal to the desired cumulative spectrum $\mathfrak{S}_X(f)$.

We will assume that the spectrum of $X(n)$ is (absolutely) continuous, that is, it has a power spectrum density $S_X(f)$ such that

$$C_{\mathcal{W}}(f) = \mathfrak{S}_X(f) = \int_0^f S_X(g)\,dg. \tag{9.5.3}$$

For computational purposes the random integral (9.5.2) has to be discretized. More precisely, we have to chose an integer K, and partition

$$f_0 = 0, \qquad f_1 = \frac{1}{K}, \qquad f_2 = \frac{2}{K}, \qquad \ldots, \qquad f_{K-1} = \frac{1}{K-1}, \qquad f_K = 1,$$

of the interval $[0, 1]$, and replace the right-hand side of (9.5.2) by the sums

$$X(n) = \sum_{k=1}^{K} e^{j2\pi nf_k}(\mathcal{W}(f_k) - \mathcal{W}(f_{k-1}))$$
$$= \sum_{k=1}^{K} e^{j2\pi n(k/K)}\left(\mathcal{W}\left(\frac{k}{K}\right) - \mathcal{W}\left(\frac{k-1}{K}\right)\right).$$

The increments

$$\mathcal{W}\left(\frac{1}{K}\right)-\mathcal{W}\left(\frac{0}{K}\right),\ \mathcal{W}\left(\frac{2}{K}\right)-\mathcal{W}\left(\frac{1}{K}\right),\ \ldots,\ \mathcal{W}\left(\frac{K}{K}\right)-\mathcal{W}\left(\frac{k-1}{K}\right)$$

are zero-mean, uncorrelated and have, respectively, variances

$$\sigma_1^2=\int_0^{1/K} S_X(f)\,df,$$

$$\sigma_2^2=\int_{1/K}^{2/K} S_X(f)\,df,\quad \ldots,$$

$$\sigma_K^2=\int_{(K-1)/K}^{1} S_X(f)\,df.$$

Thus the simulation algorithm calls for the following steps:

Step 0. Select a positive integer K determining the accuracy of our simulation.

Step 1. Generate, via a random number generator, a sequence

$$\xi_1,\xi_2,\ldots,\xi_K$$

of zero-mean, variance one, uncorrelated random values of an otherwise arbitrary distribution.

Step 2. Calculate variances

$$\sigma_1^2,\sigma_2^2,\ldots,\sigma_K^2$$

defined above via the desired power spectrum density.

Step 3. Calculate numbers

$$x_n=\sum_{k=1}^{K} e^{j2\pi n(k/K)}\sigma_k\xi_k,\quad n=1,2,\ldots,N.$$

They represent an approximate sample of our desired random signal.

Step 4. Plot the sequence x_n, $n=1,2,\ldots,N$, as a function of variable n.

Remark 9.5.1. It should be observed that if the power spectrum density is symmetric about the midpoint $f=\frac{1}{2}$, that is, $S_X(\frac{1}{2}+f)=S_X(\frac{1}{2}-f)$, then the autocorrelation function is real-valued because

$$\gamma_X(n)=\int_0^1 e^{j2\pi nf}S_X(f)\,df=\int_0^1 \cos(2\pi nf)S_X(f)\,df.$$

In those cases, we will just simulate the real parts of the sequence x_n, that is, the sequence

$$\operatorname{Re}x_n=\sum_{k=1}^{K}\cos\left(2\pi n\left(\frac{k}{K}\right)\right)\sigma_k\xi_k,\quad n=1,2,\ldots,N.$$

We shall illustrate the above algorithm on a concrete example implemented in the symbolic manipulation language *Mathematica*.

Example 9.5.1. We shall simulate a discrete-time signal $X(n)$, $n = 1, 2, \ldots, 150$, with the spectral density function $S_X(f) = f(1-f)$, $0 \le f \le 1$, pictured in Figure 9.5.1.

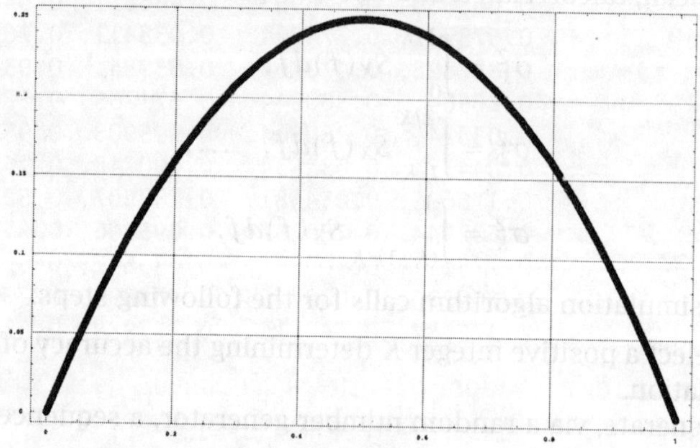

Fig. 9.5.1.

Step 0. Select a positive integer K determining the accuracy of the simulation.

```
In[1] := K= 100

Out[1] = 100
```

Step 1. Generate, via a random number generator, a sequence

$$\xi_1, \xi_2, \ldots, \xi_K,$$

of zero-mean, variance one, uncorrelated random values of an otherwise arbitrary distribution. Here we start with a sample $\eta_1, \ldots, \eta_{100}$ of 100 random numbers uniformly distributed on the interval $[0, 1]$.

```
In[2] := Table[Random[Real,{0,1}],{100}]

Out[2] = {0.85403,   0.4953,     0.87823,   0.77297,
          0.28679,   0.73909,    0.43157,   0.86123,
          0.5221,    0.34007,    0.35229,   0.71826,
          0.42662,   0.75551,    0.49997,   0.25472,
          0.34842,   0.49913,    0.44984,   0.044822,
          0.18391,   0.63281,    0.59046,   0.48641,
          0.32988,   0.1373,     0.71224,   0.71345,
          0.043092,  0.39825,    0.280669,  0.85222,
          0.52095,   0.058176,   0.92837,   0.133959,
          0.094325,  0.30266,    0.428401,  0.879233,
```

```
            0.745903, 0.80353,  0.978561, 0.834411,
            0.561987, 0.17072,  0.388094, 0.347992,
            0.2321,   0.033379, 0.675853, 0.634539,
            0.189008, 0.63513,  0.395184, 0.782319,
            0.668056, 0.576953, 0.466811, 0.64836,
            0.573731, 0.274294, 0.038411, 0.769127,
            0.827828, 0.470764, 0.05985,  0.934716,
            0.26584,  0.300044, 0.671756, 0.586724,
            0.03374,  0.266664, 0.995903, 0.952186,
            0.844731, 0.631534, 0.600718, 0.169866,
            0.176675, 0.054581, 0.133907, 0.521506,
            0.602944, 0.780287, 0.095496, 0.75238,
            0.775117, 0.309523, 0.035647, 0.817664,
            0.509276, 0.0094794, 0.363891, 0.230939,
            0.475537, 0.742815, 0.367988, 0.278754}
```

The uniform random quantity η has mean $\frac{1}{2}$ and variance $\frac{1}{12}$. Using the standard normalization formula

$$\xi = \frac{\eta - \frac{1}{2}}{\sqrt{\frac{1}{12}}},$$

we obtain a zero-mean, variance one, sample $\xi_1, \ldots, \xi_{100}$.

```
In[3] := xi = Table[(eta[[k]]-1/2)/Sqrt[1/12],
                  {k,1,100}]

Out[3] = { 1.22639,   -0.0156935,  1.31022,
           0.945582,  -0.738565,   0.828234,
          -0.237045,   1.25135,    0.0766943,
          -0.554004,  -0.51166,    0.756076,
          -0.254171,   0.885137,  -0.0000963994,
          -0.849654,  -0.525082,  -0.00301462,
          -0.17376,   -1.57678,   -1.09495,
           0.460067,   0.313388,  -0.0470465,
          -0.589289,  -1.25629,    0.735221,
           0.739423,  -1.58278,   -0.352474,
          -0.759784,   1.22012,    0.0725808,
          -1.53052,    1.48393,   -1.268,
          -1.4053,    -0.683607,  -0.248027,
           1.3137,     0.851834,   1.05146,
           1.65778,    1.15843,    0.21473,
          -1.14066,   -0.387655,  -0.526572,
          -0.928031,  -1.61642,    0.609174,
           0.466056,  -1.07731,    0.468105,
          -0.363092,   0.977983,   0.582164,
           0.266574,  -0.114969,   0.513934,
           0.255412,  -0.78187,   -1.59899,
           0.932282,   1.13563,   -0.101277,
```

```
-1.52473,   1.5059,    -0.811153,
-0.692669,  0.594981,   0.300422,
-1.61517,  -0.808299,   1.71786,
 1.56642,   1.19418,    0.455647,
 0.348899, -1.14362,   -1.12003,
-1.54298,  -1.26818,    0.0745002,
 0.356609,  0.970942,  -1.40124,
 0.874269,  0.953032,  -0.659831,
-1.60857,   1.10042,    0.0321342,
-1.69921,  -0.471497,  -0.932054,
-0.084744,  0.841137,  -0.457303,
-0.76642}
```

Step 2. Calculate the standard deviations,

$$\sigma_1, \sigma_2, \ldots, \sigma_K,$$

defined via the above power spectrum density.

```
In[4] := SX[f_] := f*(1-f)

In[5] := sigma = Table[Sqrt[NIntegrate[SX[f],
                 {f,(k-1)/100, (k)/100}]], {k,1,100}]

Out[5] = {0.00704746, 0.0121518, 0.0156098,
          0.0183757, 0.0207284, 0.0227962,
          0.0246509, 0.0263376, 0.0278867,
          0.0293201, 0.030654,  0.0319009,
          0.0330706, 0.0341711, 0.0352089,
          0.0361893, 0.0371169, 0.0379956,
          0.0388287, 0.039619,  0.0403691,
          0.0410812, 0.0417572, 0.0423989,
          0.0430078, 0.0435852, 0.0441324,
          0.0446505, 0.0451405, 0.0456034,
          0.0460398, 0.0464507, 0.0468366,
          0.0471982, 0.047536,  0.0478505,
          0.0481422, 0.0484114, 0.0486587,
          0.0488842, 0.0490884, 0.0492714,
          0.0494335, 0.0495749, 0.0496957,
          0.0497963, 0.0498765, 0.0499366,
          0.0499767, 0.0499967, 0.0499967,
          0.0499767, 0.0499366, 0.0498765,
          0.0497963, 0.0496957, 0.0495749,
          0.0494335, 0.0492714, 0.0490884,
          0.0488842, 0.0486587, 0.0484114,
          0.0481422, 0.0478505, 0.047536,
          0.0471982, 0.0468366, 0.0464507,
          0.0460398, 0.0456034, 0.0451405,
          0.0446505, 0.0441324, 0.0435852,
          0.0430078, 0.0423989, 0.0417572,
```

```
0.0410812,  0.0403691, 0.039619,
0.0388287,  0.0379956, 0.0371169,
0.0361893,  0.0352089, 0.0341711,
0.0330706,  0.0319009, 0.030654,
0.0293201,  0.0278867, 0.0263376,
0.0246509,  0.0227962, 0.0207284,
0.0183757,  0.0156098, 0.0121518,
0.00704746}
```

Step 3. Calculate the numbers

$$\operatorname{Re} x_n = \sum_{k=1}^{K} \cos\left(2\pi n \frac{k}{K}\right) \sigma_k \xi_k, \quad n = 1, 2, \ldots, N,$$

for $N = 150$. They represent an approximate sample of our desired random signal.

```
In[6] ReXi = Table[N[Sum[Cos[2*Pi*n*(k/100)]
              * sigma[[k]] * xi[[k]], {k,1,100}]],
             {n,1,150}]

Out[6] = {-0.0744418,    0.204584,    0.134502,
          -0.0940191,    0.231016,   -0.698734,
           0.506229,    -0.0506985,   0.629228,
          -0.316749,    -0.614843,    0.409999,
           0.0202642,    0.281374,   -0.137281,
          -0.239042,     0.1893,      0.00635446,
          -0.370783,    -0.117887,   -0.0400732,
          -0.0451926,    0.110532,   -0.701752,
           0.728623,     0.0365847,  -0.132731,
           0.581089,     0.0573342,  -0.723034,
           0.0928521,   -0.061326,   -0.129562,
           0.106508,     0.144272,   -0.154514,
           0.531261,    -0.270855,   -0.201814,
           0.159987,     0.0433187,  -0.206561,
           0.329139,    -0.0594249,  -0.263336,
          -0.251674,     0.0751697,   0.00108184,
           0.027182,     0.36114,     0.027182,
           0.00108184,   0.0751697,  -0.251674,
          -0.263336,    -0.0594249,   0.329139,
          -0.206561,     0.0433187,   0.159987,
          -0.201814,    -0.270855,    0.531261,
          -0.154514,     0.144272,    0.106508,
          -0.129562,    -0.061326,    0.0928521,
          -0.723034,     0.0573342,   0.581089,
          -0.132731,     0.0365847,   0.728623,
          -0.701752,     0.110532,   -0.0451926,
          -0.0400732,   -0.117887,   -0.370783,
           0.00635446,   0.1893,     -0.239042,
```

```
-0.137281,   0.281374,    0.0202642,
 0.409999,  -0.614843,   -0.316749,
 0.629228,  -0.0506985,   0.506229,
-0.698734,   0.231016,   -0.0940191,
 0.134502,   0.204584,   -0.0744418,
-0.264185,  -0.0744418,   0.204584,
 0.134502,  -0.0940191,   0.231016,
-0.698734,   0.506229,   -0.0506985,
 0.629228,  -0.316749,   -0.614843,
 0.409999,   0.0202642,   0.281374,
-0.137281,  -0.239042,    0.1893,
 0.00635446, -0.370783,  -0.117887,
-0.0400732, -0.0451926,   0.110532,
-0.701752,   0.728623,    0.0365847,
-0.132731,   0.581089,    0.0573342,
-0.723034,   0.0928521,  -0.061326,
-0.129562,   0.106508,    0.144272,
-0.154514,   0.531261,   -0.270855,
-0.201814,   0.159987,    0.0433187,
-0.206561,   0.329139,   -0.0594249,
-0.263336,  -0.251674,    0.0751697,
 0.00108184, 0.027182,    0.36114}
```

Step 4. Plot the sequence x_n as a function of variable n (see Figure 9.5.2). The consecutive values $x_1, \ldots, x_{150}$ were joined to better show their progression in time.

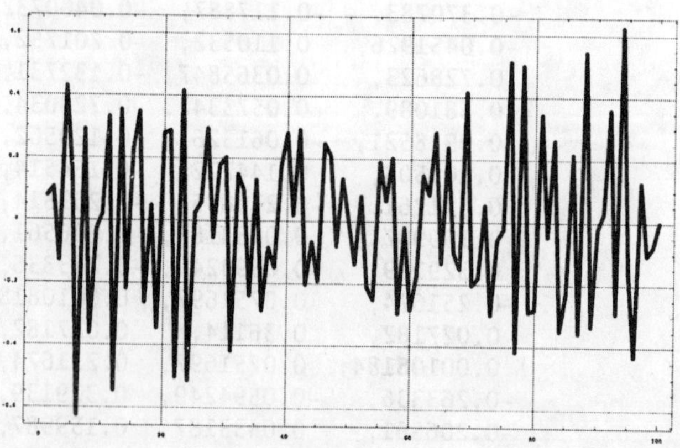

Fig. 9.5.2.

```
In[7] := ListPlot[ReXi, PlotJoined -> True,
         Frame -> True, GridLines -> Automatic]

Out[7] -Graphics-
```

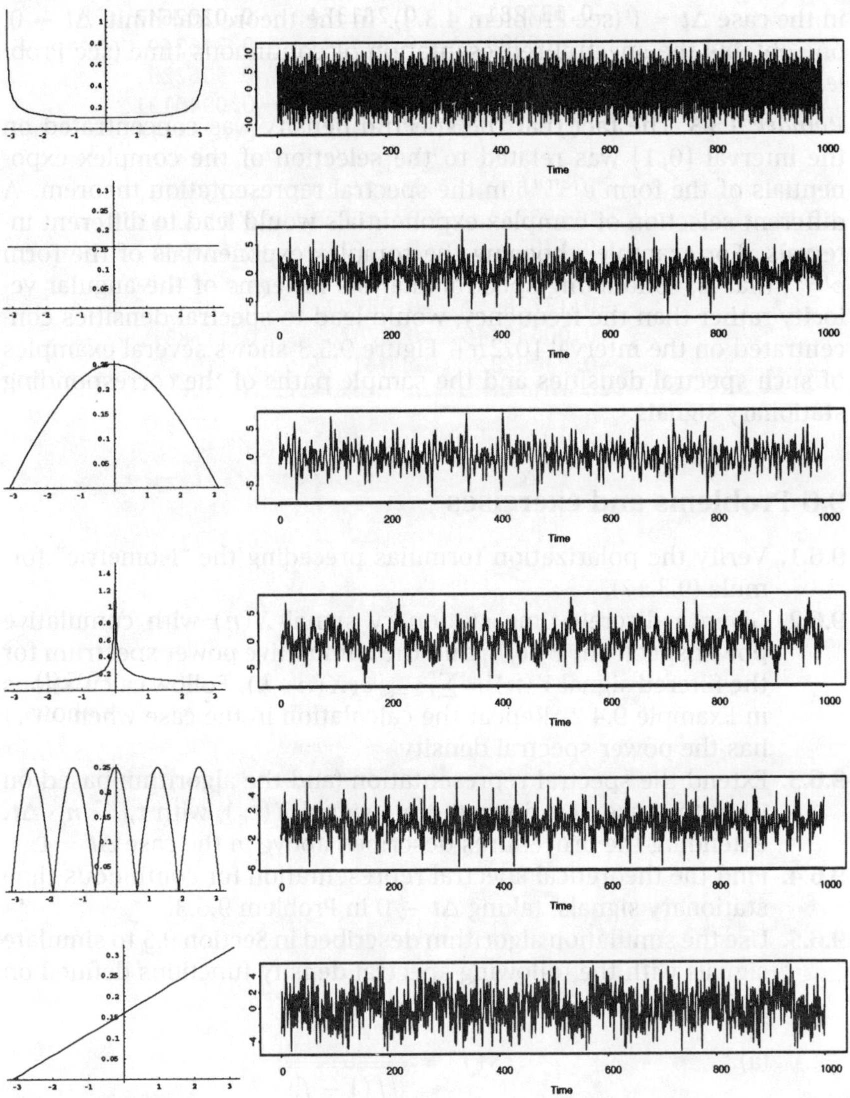

Fig. 9.5.3. Examples of simulated discrete-time stationary signals (right column) with prescribed spectral density functions (left column). Note that the spectral densities in these simulations are even and concentrated on the interval $-2\pi \le \omega \le 2\pi$.

Note that, for $K = 100$, the smallest frequency present in the representation is $f = \frac{1}{100}$. Thus the produced signal sample is periodic with period $P = 100$.

Remark 9.5.2. The above simulation can be adapted to any dicrete-time signal $X(t_n)$ with $t_n = n \cdot \Delta t$, extending the procedures described above in the case $\Delta t = 1$ (see Problem 4.3.3). In the theoretical limit, $\Delta t \to 0$, one obtains the spectral representation of continuous time (see Problem 4.3.4).

Remark 9.5.3. The fact that the spectral density was concentrated on the interval $[0, 1]$ was related to the selection of the complex exponentials of the form $e^{j2\pi nf}$ in the spectral representation theorem. A different selection of complex exponentials would lead to different intervals. For example, choosing the complex exponentials of the form $e^{jn\omega}$, that is, conducting spectral analysis in terms of the angular velocity rather than the frequency, would lead to spectral densities concentrated on the interval $[0, 2\pi]$. Figure 9.5.3 shows several examples of such spectral densities and the sample paths of the corresponding stationary signals.

9.6 Problems and exercises

9.6.1. Verify the polarization formulas preceding the "isometric" formula (9.3.17).

9.6.2. Given a discrete-time stationary signal $X(n)$ with cumulative power spectrum $\mathfrak{S}_X(f)$, find the cumulative power spectrum for the filtered signal $Y(n) = \sum_{k=-\infty}^{\infty} c_k X(n-k)$. Follow calculations in Example 9.4.2. Repeat the calculation in the case when $X(n)$ has the power spectral density.

9.6.3. Extend the spectral representation (and the algorithm based on it) in the case of a discrete-time signal $X(t_n)$, with $t_n = n \cdot \Delta t$, extending the procedures described above in the case $\Delta t = 1$.

9.6.4. Find the theoretical spectral representation for continuous-time stationary signals, taking $\Delta t \to 0$ in Problem 9.6.3.

9.6.5. Use the simulation algorithm described in Section 9.5 to simulate signals with the following spectral density functions defined on the interval $0 \le f \le 1$.

(a) $$S(f) = \frac{1}{\sqrt{f(1-f)}},$$

(b) $$S(f) = \frac{2}{3},$$

(c) $$S(f) = \cos(\pi f),$$

(d) $$S(f) = 1 - f,$$

(e) $$S(f) = |\sin(8\pi f)|.$$

9.6.6. Produce plots of several sample paths of the cumulative discrete-time white noise defined in Section 9.3.

9.6.7. Verify that the additivity property (9.3.7) of any function forces its linear form (9.3.8).

Bibliographical Comments

The classic modern treatise on the theory of Fourier series and integrals which influenced much of the harmonic analysis research in the second half of the twentieth century is

[1] A. Zygmund, *Trigonometric Series*, Cambridge University Press, Cambridge, UK, 1959.

More modest in scope, but perhaps also more usable for the intended reader of this text, are

[2] H. Dym and H. McKean, *Fourier Series and Integrals*, Academic Press, New York, 1972,

[3] T. W. Körner, *Fourier Analysis*, Cambridge University Press, Cambridge, UK, 1988,

[4] E. M. Stein and R. Shakarchi, *Fourier Analysis: An Introduction*, Princeton University Press, Princeton, NJ, 2003,

[5] P. P. G. Dyke, *An Introduction to Laplace Transforms and Fourier Series*, Springer-Verlag, New York, 1991.

The above four books are now available in paperback.

The Schwartz distributions (generalized functions), such as the Dirac delta impulse and its derivatives, with special emphasis on their applications in engineering and the physical sciences, are explained in

[6] F. Constantinescu, *Distributions and Their Applications in Physics*, Pergamonn Press, Oxford, UK, 1980,

[7] T. Schucker, *Distributions, Fourier Transforms and Some of Their Applications to Physics*, World Scientific, Singapore, 1991,

[8] A. I. Saichev and W. A. Woyczyński, *Distributions in the Physical and Engineering Sciences, Vol. 1: Distributional and Fractal Calculus, Integral Transforms, and Wavelets*, Birkhäuser Boston, Cambridge, MA, 1997.

The second volume of the above text,

[9] A. I. Saichev and W. A. Woyczynski, *Distributions in the Physical and Engineering Sciences, Vol. 2: Linear, Nonlinear, Fractal, and Random Dynamics in Continuous Media*, Birkhäuser Boston, Cambridge, MA, 2005,

is scheduled to appear later this year.

Good elementary introductions to probability theory, and accessible reads for the engineering and the physical sciences audience, are

[11] S. M. Ross, *Introduction to Probability Models*, Academic Press, Burlington, MA, 2003,

[10] J. Pitman, *Probability*, Springer-Verlag, New York, 1993.

On the other hand,

[12] M. Denker and W. A. Woyczyński, *Introductory Statistics and Random Phenomena: Uncertainty, Complexity, and Chaotic Behavior in Engineering and Science*, Birkhäuser Boston, Cambridge, MA, 1998

deals with a broader issue of how randomness appears in diverse models of natural phenomena and with the fundamental question of the meaning of randomness itself.

More ambitious, mathematically rigorous treatments of probability theory, based on measure theory, can be found in

[13] P. Billingsley, *Probability and Measure*, Wiley, New York, 1983,

[14] O. Kallenberg, *Foundations of Modern Probability*, Springer-Verlag, New York, 1997,

[15] M. Loève, *Probability Theory*, Van Nostrand, Princeton, NJ, 1961.

All three also contain a substantial account of the theory of stochastic processes.

Readers more interested in the general issues of statistical inference and, in particular, parametric estimation, should consult

[16] G. Casella and R. L. Berger, *Statistical Inference*, Duxbury, Pacific Grove, CA, 2002

or

[17] D. C. Montgomery and G. C. Runger, *Applied Statistics and Probability for Engineers*, Wiley, New York, 1994.

The classic texts on the general theory of stationary processes (signals) are

[18] H. Cramer and M. R. Leadbetter, *Stationary and Related Stochastic Processes: Sample Function Properties and Their Applications*, Dover Books, New York, 2004,

[19] A. M. Yaglom, *Correlation Thoery of Stationary and Related Random Functions*, Vols. I and II, Springer-Verlag, New York, 1987.

However, the original,

[20] N. Wiener, *Extrapolation, Interpolation, and Smoothing of Stationary Time Series*, MIT Press and Wiley, New York, 1950,

still reads very well.

Statistical tools in the spectral analysis of stationary discrete-time random signals (also known as *time series*) are explored in

[21] P. Bloomfield, *Fourier Analysis of Time Series: An Introduction*, Wiley, New York, 1976,

[22] P. J. Brockwell and R. A. Davis, *Time Series: Theory and Methods*, Springer-Verlag, New York, 1991,

and difficult issues of analysis of nonlinear and nonstationary random signals are tackled in

[23] M. B. Priestley, *Non-Linear and Non-Stationary Time Series Analysis*, Academic Press, London, 1988,

[24] W. J. Fitzgerald, R. L. Smith, A. T. Walden, and P. C. Young, eds., *Nonlinear and Nonstationary Signal Processing*, Cambridge University Press, Cambridge, UK, 2000.

The latter is a collection of articles, by different authors, on the current research issues in the area.

A more engineering approach to random signal analysis can be found in a large number of sources, including

[25] A. Papoulis, *Signal Analysis*, McGraw-Hill, New York, 1977,

[26] R. G. Brown and P. Y. Hwang, *Introduction to Random Signal Analysis and Kalman Filtering*, Wiley, New York, 1992.

A general discussion of transmission of signals through linear systems can be found in

[27] M. J. Roberts, *Signals and Systems: Analysis of Signals Through Linear Systems*, McGraw-Hill, New York, 2003,

[28] B. D. O. Anderson and J. B. Moore, *Optimal Filtering*, Dover Books, New York, 2005.

Gaussian stochastic processes are thoroughly investigated in

[29] I. A. Ibragimov and Yu. A. Rozanov, *Gaussian Random Processes*, Springer-Verlag, New York, 1978,

[30] M. A. Lifshits, *Gaussian Random Functions*, Kluwer Academic Publishers, Dordrecht, The Netherlands, 1995,

and for a review of the modern mathematical theory of not necessarily second-order and not necessarily Gaussian stochastic integrals, we refer to

[31] S. Kwapien and W. A. Woyczyński, *Random Series and Stochastic Integrals: Single and Multiple*, Birkhäuser Boston, Cambridge, MA, 1992.

Index

Also by Wojbor A. Woyczyński

S. Kwapień and W. A. Woyczyński, *Random Series and Stochastic Integrals: Single and Multiple*, Probability and its Applications, Birkhäuser Boston, 1992, Softcover, ISBN: 0-8176-4198-X.

A. I. Saichev and W. A. Woyczyński, *Distributions in the Physical and Engineering Sciences, Volume 1: Distributional and Fractional Calculus, Integral Transforms and Wavelets*, Applied and Numerical Harmonic Analysis, Birkhäuser Boston, 1996, Hardcover, ISBN: 0-8176-3924-1.

M. Denker, W. A. Woyczyński, and B. Ycart, *Introductory Statistics and Random Phenomena: Uncertainty, Complexity and Chaotic Behavior in Engineering and Science*, Statistics for Industry and Technology, Birkhäuser Boston, 1997, Hardcover, ISBN: 0-8176-4031-2.

R. Kałuża (Translated and edited by A. Kostant and W. A. Woyczyński), *Through a Reporter's Eyes: The Life of Stefan Banach*, Birkhäuser Boston, 1996, Softcover, ISBN: 0-8176-4371-0.